A *K*-means Approach to the Analysis of Spatial Structure in Upper Paleolithic Habitation Sites

Le Flageolet I and Pincevent Section 36

Jan F. Simek

BAR International Series 205
1984

B.A.R.

5, Centremead, Osney Mead, Oxford OX2 0ES, England.

B.A.R.-S205 : 'A K-means Approach to the Analysis of Spatial Structure in Upper Paleolithic Habitation Sites'

Price £ 17.00 post free throughout the world. Payments made in dollars must be calculated at the current rate of exchange and $3.00 added to cover exchange charges. Cheques should be made payable to B.A.R. and sent to the above address.

ISBN 0 86054 263 7

For details of all B.A.R. publications in print please write to the above address. Information on new titles is sent regularly on request, with no obligation to purchase.

Volumes are distributed from the publisher. All B.A.R. prices are inclusive of postage by surface mail anywhere in the world.

Printed in Great Britain

For Jean-Philippe Rigaud

ACKNOWLEDGEMENTS

Trying to express my appreciation to all of those people who have helped to make this volume is truely an impossible task. Over the years, literally hundreds of individuals, many of whom I never met, contributed their time and effort towards obtaining the data I analyzed. There are certain individuals who must be acknowledged. First, I thank Jean-Philippe Rigaud for all his time, support, and encouragement over the years. Jean-Philippe has taught me virtually all of the very little I know about Paleolithic archaeology. He showed this American how sites *really* should be excavated, gave me data and the confidence to work on it, and provided me with space in his laboratories to carry out analyses. I can only hope I did half the job he wanted done after his fifteen years of work on Le Flageolet.

Margaret Conkey served as my advisor and my friend throughout my graduate career. It was Meg who first interested me in problems paleolithic and who supported my choice of direction and goals. Meg knows how much she has meant to the completion of this work.

Where Rigaud provided me with the empirical base for my research, Albert Ammerman showed me how to confront the problems I addressed and pointed out theoretical and methodological paths that opened the way for my research. This work is as much Albert's as it is mine.

I must also thank Richard Beerbower, John Fritz, Donald Grayson, Michael Little, and Charles Redman, all of whom read and commented on various draft manuscripts. Some of these people contributed as members of my dissertation committee while others simply read my work because I asked them to.

Very special thanks and admiration are due Keith Kintigh, who wrote the *k*-means program used in the studies presented here and made sure that it worked for me. Keith's implementation represents a great stride forward for archaeological analysis. Without his efforts, this research would never have been possible.

There are many in France who helped me with my work and simply enriched my life with their presence. I thank Jean-Michel Geneste for his help, knowledge and constant wit, and I thank his family, Fabienne and Lea, for taking in a lonely amerloque. In Rigaud's office, Monique Sigaud was always helpful with her time, energy and knowledge. Cathy Saysset helped me to learn french and became one of my greatest friends. Jeanne Hennebois also taught me the language, admittedly with a distinctive accent, and made many of the years at "La Beynaque" pleasureable.

Jean-George Marcillaud drew the stone tools that illustrate this study with the hand of a master, but that is just a little of what he did for me. I hope he will always be my friend.

At the University of Washington, Robert C. Dunnell has been a constant source of support, encouragement, and understanding. It's nice to be treated with kid gloves on one's first job.

Several people provided important help at various times during my education and research. Susan Simek and Charlotte Beck helped to type the final manuscript. Mike Zisser made reduced photograhs of the illustrations from large-sized originals. I also thank Dr. Anthony R. Hands of B. A. R. for his help during preparation of the final copy.

Like many of the people listed above, Roy Larick has been with me during the entire research period. Unlike any of the others, Roy shared the whole research experience: living in foreign places, fighting the urge to quit and go home, exploring ideas and areas never dealt with before. It doesn't seem reasonable to thank Roy for something we're only in the middle of. Nevertheless, I acknowledge his aid over the last nine years.

And to Robin, who has been beside me all along even though it's been more than difficult, I can only offer my love and apologies for all the time I've spent with this instead of with her.

TABLE OF CONTENTS

LIST OF TABLES

LIST OF FIGURES

CHAPTER 1

GENERAL INTRODUCTION

THE CONTEXT OF THE PROBLEM

The research presented in the following dissertation began with the explicit goal of identifying different areas within a series of Upper Paleolithic habitation sites where the prehistoric inhabitants had carried out different specific activities. This goal is consistent with a long-standing interest in paleolithic archaeology for reconstructing the occupations at sites to allow their integration into a broader, regional-scale understanding of past human adaptations and behavioural change (e.g., Binford and Binford 1966). The approach taken here involves the analysis of artifact distributions in space, or "spatial analysis", since it seems reasonable to suppose that the locations of artifacts in a site will be influenced by the various tasks performed in different areas (Whallon 1973, 1974; Dacey 1973) and will therefore reflect certain aspects of individual site use.

In the studies that follow, two Aurignacian occupation levels from a rockshelter site in southwestern France called Le Flageolet are analyzed. The dissertation's focus on spatial analysis is also due to the specific problems posed by the Le Flageolet data and to the excavation techniques used there. The levels yielded thousands of bone and stone artifacts found scattered through a jumble of very large rocks that fell from the shelter roof during the Pleistocene. As each of these artifacts was dug from the ground, its position in space was recorded by three dimensional coordinates. This form of recovery provided a data set particularly suited to spatial analysis. However, the vast number of artifacts cataloged in this way and the configuration of the rockshelter itself made the analysis of spatial patterns in the artifact distributions difficult. To solve these problems, mathematical techniques for analyzing spatial distributions were sought and applied. These studies allow the definition of spatial patterns in the artifact distributions, as was the initial goal of the research. Yet, as will be seen, the <u>interpretation</u> of the patterns was not possible in the anticipated framework--as reflecting the spatial locations of specific past human activities.

The Activity Model for Spatial Process

Archaeologists have long assumed that the distribution of artifacts across an excavated site surface holds information concerning the activities performed at the site by human inhabitants. In the traditional view, activities are rather specific kinds of human behaviour. They are discrete in that they involve one or a few actions, require only a few implements, and have a single, usually functional goal (cf. Schiffer 1976: 66-70; Binford 1983: 147). Activities also have spatial integrity; they are carried out within a limited area. Hide-scraping is an example of such an activity. Typically, hides are staked out on the ground in open areas and are scraped with rather specific tools (e.g., scrapers or fleshers). Traditionally, archaeologists have viewed sites as collections of these kinds of "activity areas". Thus, the task of spatial analysis has been the definition of clusters of objects over space that reflect these areas. An outgrowth of this approach is a concern with identifying tool kits over space. Tool kits are the artifacts used to perform a task and are defined, once activity areas (clusters) have been located, by counting up the different kinds of artifacts in each spatial cluster and looking for "cooccurring" sets of objects. The result of such an approach is usually a site map with activity areas outlined and a list of the tool kits used in different activities (e.g., Van Noten, et al. 1979). While such reconstructions are attractive, the assumptions underlying this approach are probably too simple.

In recent years, information derived from a variety of research contexts, most notably from ethnoarchaeology, has made archaeologists aware that the activity model may have serious flaws (e.g., Clarke 1973, 1977; Clark 1979; Gould 1977). John Yellen's observations at modern !Kung San sites have been particularly provocative in this regard (1977). Yellen shows that a complex set of human behaviours acts to eliminate the kind of artifact patterning expected by the activity model. For example, the number of social groups present during an occupation may affect the nature and extent of garbage disposal and the spatial organization of refuse deposits (cf. Binford 1983: 138-42). Yellen's work will be discussed in more detail in Chapter 2.

Ethnoarchaeology has led to re-evaluation of the processes involved in generating distributions of objects in archaeological sites. More general questions have also arisen concerning the notion of "tool kit" as usually applied in traditional research. Tool kits are conventionally viewed as the equipment required by an individual performing a given task (Binford and Binford 1966; Binford 1983: 147). A rather important jump is made from this definition to the assumption that all this equipment will be left in place where it was used after the task is complete. In fact, this is not likely to be the case due to the nature of artifact deposition even before such things as refuse disposal occur. For example, Ammerman and Feldman (1974) argue that the probability of a given artifact being deposited into the archaeological record at a given location was influenced by a number of processes, including loss, breakage, frequency of activity performance, the number of different uses to which a given object was put, and the value of the object in terms of raw material and labor invested in production. They show that artifact loss rates conform to a random probability model. For any particular object used in any specific activity event, the probability of deposition through loss is quite low and, for the most part, unrelated to the activity. A multinomial sampling distribution seems to provide the best model for object deposition in this mode. Because of this, the archaeological assemblage left in a place where a set of tools was used in a task has very little chance of reflecting the kit accurately; during a single task performance, relatively few objects will be lost, broken, or worn out, and the frequencies of these by tool classes will be determined at random. In short, there is little chance of finding archaeological assemblages representative of tool kits actually "mapped" in their location of use.

Another process adding to doubts about activity area integrity is what Binford has called "curation" (Binford 1973, 1976, 1979b). This is a rather simple concept that really encompasses a series of processes whereby artifacts are conserved for further use. Curation implies that artifacts not lost, broken, or worn out are unlikely to be deposited in the use area because they are still useful. Such objects might be stored elsewhere in the site, reworked into other tools, or even removed entirely from a site upon abandonment. Thus, curation processes may affect the expected correspondance between activity areas and tool kits. It is becoming clear that when human behaviour with "things" is considered in all i complexity, the assumptions underlying the traditional "activity area" approach to spatial analysis are unwarranted.

Research findings such as these have encouraged a general re-evaluation of spatial analysis in archaeology. Perhaps most importantly, it is now acknowledged that artifact distributions are the results of formation processes rather than simple "frozen pictures of the past." A change in perception is occurring; archaeologists are beginning to talk about spatial process as the goal of spatial analysis and to view material distributions as the results of dynamic forces that may be quite varied in nature and over time (Clarke 1977). Complete conceptual change has yet to be fully realized, but it represents one of the exciting frontiers of modern archaeological research.

This change is clearly manifested in the research presented here. The studies in this volume were initiated under the activity area model. I set out, at first, to find clusters of objects in the occupation surfaces from Le Flageolet that would allow me to identify tool kits. During the course of research, analytic results were not consistent with the activity area model. Further work, both analysis and with the archaeological literature, indicated that a more complex view of spatial distributions was required, one accounting for the diverse processes that formed the archaeological record rather than assuming a single inferential model. This is, perhaps, as research should be--confusing until unwarranted assumptions are eliminated. However, the process of research influenced the final product to a great extent. It therefore seems profitable to trace the development of the studies presented here.

CHANGING RESEARCH GOALS

My research began in 1978 after three years of fieldwork under Jean-Philippe Rigaud at Le Flageolet had stimulated my interest in the spatial analysis of archaeological distributions. Rigaud's work on the distributions from Couche VII at Le Flageolet shows that different kinds of fractures in reindeer bone have distinct spatial patterns (Delpech and Rigaud 1974; Rigaud 1978). He found areas within the site where bone refuse had been deposited after being rendered for grease, areas where primary butchering refuse had been piled, and areas where bone debris reflected marrow cracking more than other processing techniques. However, Rigaud's analyses were time-consuming and demanding, since they required the intensive visual inspection of complex distribution maps. The analysis of many more artifact classes and the comparison of patterns between them clearly calls for an automated or computer-assisted technique. My work began as a search for quantitative methods that could detect patterning within and between many artifact classes. With these methods, I wanted to identify areas within the site, analogous to those defined by Rigaud based on bone data, where hopefully several stone tool classes had been used and deposited together.

In undertaking an analysis of activity areas at Le Flageolet, the first order of business involved a search for an analytic method that would accomodate the peculiar data from the site's Aurignacian levels. These data had been resistant to quantitative treatment for some time because of the "naturally clustered" configuration produced by large rocks that fell from the roof of the shelter during the Pleistocene. Most of the methods for spatial analysis that have been applied to archaeological data involve the definition of clusters of objects in artifact distributions. This approach makes sense in terms of the activity focus outlined above. However, in dealing with point maps, clusters represent areas where points are concentrated, i.e., bounded by more-or-less empty spaces. The large roof-fall blocks in Le Flageolet act as empty areas within the distribution since most of the cultural material was deposited in the cracks around them rather than over their surfaces. Traditional analytic techniques easily found the within-crack clusters. Yet, to understand the artifact distributions it was necessary to go beyond this configuration to find patterns (1) over larger areas of the site and (2) within the material distributions contained in a crack between rocks. Otherwise, spatial analysis would provide nothing more than could be achieved by inspecting the map of rocks.

A method that would accomplish such analysis was found. In 1978, Albert Ammerman introduced me to a new approach that he and Keith Kintigh were developing called "heuristic spatial analysis" (Kintigh and Ammerman 1982). At the base of their approach is an expanded _k_-means clustering algorithm (cf. Hodson 1970) that seemed to avoid many of the pitfalls inherent in other technqiues. Roy Larick, a fellow graduate student at S.U.N.Y., Binghamton, and I became involved in refining the _k_-means technique, anticipating its use for the Le Flageolet data. As an initial application, we analyzed artifact distributions from another Upper Paleolithic site, Pincevent Section 36 (Leroi-Gourhan and Brezillon 1972), that did not have the "preclustered" configuration produced by rocks at Le Flageolet. Pincevent 36 is a well-preserved Magdalenian habitation site where excavations uncovered a series of discrete material deposits interpreted as hut structures by the analysts (_ibid_). Because of the apparant spatial integrity of the distributions, we felt that Pincevent would provide a test for the analytic method and surely reveal the expected kind of activity patterning in space. Initial analyses were encouraging. _K_-means clustering found several levels of patterning in the Pincevent distributions, and these showed coherent areas of artifact concentration. However, in the course of investigating artifact association (or the lack of it) in the discovered spatial patterns, we found that relations among the various classes often reflected the overall abundance of other materials at a given location rather than the kind of limited inter-class co-occurrence predicted by the activity model (Simek and Larick 1983). This was not only confusing but also uninterpretable in terms of tool kits (since nearly every "area" defined by spatial analysis contained a complex assortment of all artifact classes). Despite the excellent preservation at Pincevent, the kinds of artifact relations expected under the activity model--such as association among a few stone tool classes in certain areas of the site--were not found.

The results of our Pincevent analyses suggested that, while the quantitative technique performed admirably and promised to be useful for analyzing the more-complex Le Flageolet data, the interpretive model was too simple. In short, it was apparant that the kinds of processes noted by Yellen and others might be operating on the Upper Paleolithic distributions under study but that we did not have the conceptual "machinery" to distinguish them. While exciting from a research standpoint, this created some important analytic and interpretive problems.

By 1979, the need for a shift in problem orientation had become evident. Rather than a search for activity areas within artifact distributions, the study became a search for methods and models for determining _if_ activity areas were, in fact, present or if _other_ factors, such as garbage disposal, had produced the observed patterning. At the same time, the analysis of the Le Flageolet Aurignacian levels proceeded as planned. As will be seen, the results of the later studies differed from Pincevent, and these differences seemed to relate to formation processes that possibly included activity area separation at Le Flageolet.

An explanation of the differences between Pincevent and Le Flageolet was necessary. To provide this, a set of models was developed, describing some of the variability that might occur in spatial distributions of artifacts resulting from complex site formation processes. These models, although _post hoc_, were formulated to illustrate possible ways that different processes might combine. Of most importance in the model development exercise was the identification of variables that accurately measured differences in distributions related to site formation processes. These variables were defined during analysis rather than before; their definition relied on results of analysis and changes in our perception of the problem. It is for this reason that the models came late in the sequence of research. It should be stressed at the outset that the formation models presented here are only first approximations of the kind of models ideally envisioned and are not considered definitive. However, as will be shown in the following pages, they do express some of the basic differences observed in spatial patterning between two sites and should provide a useful first step in the comparison of spatial process.

OUTLINE OF THE PRESENTATION

The models that will guide much of the pattern interpretation in the studies that follow will be discussed in Chapter 2, although they had their genesis at a later stage of the research. The analytic method used in studying each level is explained in Chapter 3, since the same approach and techniques were applied in consistent fashion to all distributions. Chapter 4 presents the detailed analysis of Pincevent 36 that provided first confusion, then insight into the complex issues at hand. Chapters 5 and 6 respectively contain a general introduction to the geology and archaeology of Le Flageolet and a more detailed introduction to the problems of spatial analysis for the Aurignacian levels from the site. Chapters 7 and 8 present the spatial analyses of artifact distributions from two Le Flageolet Aurignacian occupation levels. Chapter 9 is a discussion of the results of specific analysis and a comparison of spatial organization in the three archaeological levels. In addition, some general implications of these studies for spatial archaeology and directions for further research are presented in the closing chapter.

MODELLING SITE FORMATION PROCESSES

THE DEVELOPMENT OF MULTIPLE MODELS

As stated in the preceeding chapter, a set of models for spatial structure in the Pincevent and Le Flageolet distributions was developed after most analyses were complete. This modelling occurred in order to clarify observed differences among the analyzed sites that seemed to relate to basically distinct depositional characteristics. This clarification had two goals. First, models were designed to characterize differences observed between Pincevent and Le Flageolet. Second (and perhaps most important), models were constructed to relate some potential causes of spatial variation to the patterns observed in specific cases.

The process of model development involved a long dialogue between data analysis and interpretation. Thus, the models consider variables identified as important during analyses. For example, studies of the Pincevent distributions indicated that the relative abundance of materials at given locations across the site surface conditioned the relative frequencies of artifact classes present (Simek and Larick 1983). Moreover, this relation between abundance of material and artifact class frequencies varies as the multiple pattern scales recognized by k-means analysis are examined. These observations led to the consideration of "homogeneity" in cluster content over the site surface. Similarly, the search for and recognition of different levels of spatial patterning in each distribution made the inclusion of "spatial refinement" appropriate in the models. These two variables will be discussed in more detail shortly.

After differences in spatial patterns were noted, the best way to express those differences was sought. Once initial expectations for spatial relations were proved false, a shift in focus occurred. Observed patterning had to be examined in terms of multiple processes. This explanatory realm has mostly been ignored by archaeologists. Typically, models in archaeology are expressed in terms of a single process (artifact use, economic exchange, resource availability, etc.). Accounting for multiple processes is a far more difficult task than modelling a single one and is relatively undeveloped in archaeology at this point in time. Thus, when this study began, the utility of complex models was neither anticipated (especially in light of the prevailing conceptions of activity areas and tool kits) nor recognized until some analytic results were in hand. A benefit of this, however, is that the variables employed in model building account for more complex and precise dimensions of process than might otherwise have been the case.

Before discussing the model variables, however, some other concerns that were important in the research prior to its initiation, including the integration of contextual elements such as features and walls, will be considered.

CONTEXT IN THE INTERPRETATION OF SPATIAL PATTERNS

From the outset, the importance of contextual information to the analysis and interpretation of spatial patterning was recognized. Indeed, Kintigh and Ammerman (1982) stressed the integration of contextual information as basic to their "heuristic approach". That approach is an explicit attempt to provide for the integration of spatial patterning in artifact distributions with other aspects of sites that might affect the dispersal of objects (ibid: 31):

> the heuristic approach is synthetic in that it attempts to open the way for the use of contextual knowledge and human expertise within a formal (computer-executed) procedure for aiding human-directed spatial analysis

In the studies presented here, context will refer to those characteristics of a site that place constraints in any way on distributions of artifacts. Examples of contextual elements include the walls of a cave or rockshelter, the confines of a room in a house, large blocks fallen from a cave wall that break up artifact distributions (important in the Le Flageolet case), and hearths and pits that hold artifacts or serve as "centers of gravity" for deposition. Some site characteristics that provide context for artifact distributions can be and often are analyzed in spatial terms (features, for example). In such cases, other aspects of the archaeological record may serve as context for the distributions under study. Clarke (1977) argues that the different "elements" of the archaeological record must be organized hierarchically when considering context. Thus, room walls may serve as context for hearths which serve as contexts for artifacts. Clearly, context is a relative term and depends on the scale of analysis and the nature of the data elements under study (Clarke, ibid).

Context comprises those constraints beyond which spatial patterning and artifact association can be given behavioural interpretation. Conversely, patterning related to these constraints must be explained in contextual terms. For example, some of the patterning present in the Le Flageolet artifact distributions is due to the large blocks; in fact, one level of clustering can be identified that represents all the cracks between rocks. This pattern is explained in terms of the context--rocks--while patterns within and between the cracks are interpreted in other ways since they seem to be independent of this basic contextual characteristic. Contextual elements can be natural or cultural. In the studies presented here, principal contextual elements include large blocks and bedrock walls (for Le Flageolet) and varied constructed features involving the use of fire (for both Pincevent and Le Flageolet).

Various models for deposition at Pincevent and Le Flageolet are assessed within the constraints imposed by context. This means that patterning in artifact distributions are always first examined for relations to hearths, rocks, the shelter wall, etc. Of course, this kind of relationship is often recognized in archaeology. Indeed, in the traditional activity model, associations between certain activities and features are expected. These expectations are based on the assumption that certain activities--cooking food, heat treatment of chert, grease rendering--require the use of fire as a heat and light source and are thus constrained in their location to the immediate proximity of a hearth. More complex interpretive models, concerned with multiple processes of material generation and deposition, must also take context into account.

A LITTLE HELP FROM ETHNOARCHAEOLOGICAL ANALOGY

In attempting to formulate descriptive models to account for the differences observed between Pincevent and Le Flageolet, the ethnoarchaeological literature was searched for information that might inform on site formation processes and the role of contextual elements in them. It should be stressed that, while virtually any use of such information will be (by definition) analogic, no attempt will be made here to provide "tailored fits" between modern and paleolithic cases. Ethnoarchaeology provides illustrations of the potential range of variation in behaviour and context--is a source for ideas--and not the means for model verification. This discussion of ethnoarchaeological observations is designed to identify common characteristics of deposition at sites occupied by hunting peoples--in particular, any processes of deposition that are general in these kinds of sites. Do hunter-gatherers today have certain uniform ways of introducing materials into the archaeological record? To what aspects of site occupation do these common depositional processes relate? Can these be more generally recognized in the spatial organization of archaeological sites?

Two ethnoarchaeological case studies were the most influential in forming the series of models to be presented shortly. These cases were of interest because they are the only two studies devoted to the ethnoarchaeological study of spatial formation processes. The works represent observations of modern hunter-gatherer groups in quite opposite environmental situations, and they do point up some interesting general characteristics of site formation in forager settlements. One study to be considered here is Yellen's report on observations among the !Kung San of southwest Africa (1977). The second is Binford's important work among the Nunamiut Eskimo of Alaska (e.g., 1976, 1977, 1978a, 1978b, 1981a, 1981b, 1983). These case studies are emphasized because they were designed to address the problem of site formation. Binford's work in particular was designed to identify the kinds of site formation process that might operate in paleolithic contexts. Many of his observations were made with problems of the sort considered here in mind (1983: 95-108).

The !Kung San Case

Yellen's landmark monograph on !Kung San ethnoarchaeology (1977) was one of the first comprehensive actualistic studies to consider the relation between human behaviour during the occupation of a place and the spatial record remaining after abandonment. He notes that factors such as the number of people occupying a site, the social relations among them, the temporal duration of the occupation, and the reason for site use within a larger regional subsistence system all affected the kinds of materials deposited and the spatial patterns present in those deposits (ibid: 98). Indeed, Yellen argues that such factors may have a greater influence on the character of archaeological distributions than does the spatial configuration of specific activities during occupation.

!Kung San residential sites can be divided into a "nuclear" or core area centered around hearth features and "peripheral" areas away from the core (Yellen 1977: 126). The two kinds of area are characterized by quite distinct modes of artifact deposition. Core areas are the loci for a social group's domestic activities. For example, a family cooks its meals, consumes food, repairs tools, and manufactures clothing around the hearth. Sleeping occurs in a brush shelter positioned nearby. Typically, material deposits generated in the core area have a particular spatial organization--domestic activities tend to be repeated during an occupation and the space required for task performance must be kept clear of debris. Thus, refuse from individual activity events is collected and disposed of in a single "dump", usually located on the hearthside opposite the brush shelter. The archaeological record of a single family usually comprises a hearth feature and a refuse dump on one side of the hearth where refuse from many activities (related only in their context of performance) is mixed. An open area is present on the other side of the hearth where a shelter provides shade for rest, sleep, and certain group maintenance tasks.

Peripheral areas of !Kung occupation sites are the loci of activities requiring open space, involving or producing especially obnoxious materials, or needing to occupy a space for relatively prolonged periods of time. Even here, different tasks may be performed in the same location over the period of site occupation. However, peripheral area deposits tend to comprise "bouts" of specific task performances. The organization of peripheral areas can be complicated by several factors. Shelter from the sun may influence activity location. For example, material refuse from a single event of mongongo nut processing may be scattered over a larger area than required for the task as the processor moves from shaded area to shaded area during the day (Yellen 1977: 92). Overall, however, deposits created in the peripheral areas of !Kung occupation sites are more directly related to single tasks than are those in the residential core. Peripheral areas are scattered in a circular pattern around the domestic core, forming what Yellen has termed a "ring" pattern of deposits (1977: 125-31).

The presence of more than one social group in !Kung settlements also affects the organization of core and peripheral areas. When, for example, several families are present, each occupies and maintains its own domestic core area. These are located together but spaced according to (1) the amount of area required for dumping in each household and (2) the social bonds among group members (*ibid*: 128). Peripheral areas are dispersed around *all* core areas. In effect, the core represents the hearth-centered domestic areas of all families present. The rich material accumulations generated by each family at their hearth will be internally mixed in terms of activities but will be separated one from the other. The number of these core deposits is directly related to the number of families present during the occupation.

The Nunamiut Case

Lewis Binford's work among the Nunamiut Eskimo (e.g., 1978b) supports Yellen's picture of intrasite formation processes, although Binford is far more concerned with the specific processes leading to deposition than is Yellen. Binford believes that spatial process in Nunamiut sites can be understood generally in terms of (1) human body size and (2) the nature of the tasks being performed in given locations. However, he sees much material deposition resulting from three processes that are not related to activities. Two of these processes depend on the size of material being deposited. Objects are incorporated into the archaeological record by being dropped (if they are small) or tossed (if they are large). A third disposal mode involves refuse deposition or what Binford terms "aggregate" tossing. All three of these processes occur around hearths and result in hearth-area dumps, regardless of the number and relation of activities performed. In fact, disposal modes may be reflected in space while activities are not. For example, a "drop zone" made up of very small objects fallen between a worker's legs can be distinguished even when several different activities (each generating small bits) were carried out sequentially. For Binford, it is the position of the humans around the hearth that conditions the form and location of the accumulations:

> The prior existence of the small bone juice dump located behind individual no. 1 acted as a sort of 'magnet' attracting further disposal, so that most of the items tossed away by men seated at the later hearth were added to the earlier dump, still visible from the previous day [1983: 155].

In short, hearth areas are locations where many activities are performed and refuse from these ejected or dropped into mixed dump deposits.

Binford (ibid: 165) makes a distinction between "intensive" activity areas like hearths (where many activities are performed and their material results mixed) and "extensive" activity areas (that involve tasks requiring large amounts of space and are often performed in a standing position). In depositional character, Binford's extensive activity areas are Yellen's peripheral areas:

> It is generally true, in fact, that activities like roasting and butchery, which monopolize considerable amounts of space, are located away from areas used intensively on a day-to-day basis [Binford 1983: 170].
>
> Hide working is yet another activity repeatedly seen to occur in areas peripheral to the central parts of residential camps [ibid: 172].

The basic peripheral area pattern--activities separated from but situated around central domestic hearth areas--is common to Binford and Yellen's models. Also common, but not explicit in Binford's case, is the notion that activities performed in peripheral zones are less often repeated than core area tasks and thus leave more-discrete deposits in the archaeological record.

GENERAL CHARACTERISTICS OF HUNTER-GATHERER SETTLEMENT ORGANIZATION

Actualistic studies of hunter-gatherer settlements indicate general patterning in material deposition that will be used in the models presented below. Two kinds of deposit can be identified based on the work of Yellen and Binford. First, domestic dumps, representing the debris accumulated from a variety of activities, will be present in virtually every site where group maintenance tasks were carried out. Since many of these tasks are very basic--eating, for example--dumps should be almost ubiquitous. Dumps have particular contextual characteristics. Second, activity specific deposits _may_ be present in a site, but this is not universally true. These will have particular locational relations with domestic areas.

Domestic hearth deposits, or dumps, are generated by a rather specific process, namely the gradual accumulation of material through repeated bouts of refuse disposal. Most of this refuse is produced during tasks related to group maintenance. The context of these dumps may include a hearth feature, since many group maintenance tasks occur in proximity to fire. Hearth dumps are positioned near but not in the feature and are very dense on only one side of the fireplace. This is because the people generating the accumulations sat on the opposite side. This pattern characterizes hearth use in most known hunter-gatherer settlements (Binford 1983: 175; Figure 115).

Because hearth area dumps represent the refuse generated by a wide range of activities, they have particular, characteristic content. Hearth deposits may include the material byproducts of the final stages of food processing and consumption, e.g., large numbers of small, burnt bone fragments. Some tool production and repair may occur in hearth contexts, and the waste products of these activities may be incorporated in dumps. Because many tasks are performed at the hearth, a variety of tool classes broken in use or no longer servicable may be present. What needs emphasis here is the fact that hearth dumps are the results of refuse deposition _away from the actual location of activity._ Waste material _resulting from activities performed_ on one side of the hearth is most often disposed on the other side. This process forms a mixed deposit as far as activities are concerned.

The _repetition_ of various tasks in hearth contexts is also _indicated by_ ethnoarchaeological studies. This ultimately results in relatively rich accumulations occupying a confined area to one side of the hearth feature (Yellen 1977: 131). When more than one dump is present, either there were several domestic hearths in use (i.e., several groups) or contextual constraints precluded the formation of a single deposit. In either case, all hearth dumps in a site should be similar in content, since the maintenance tasks performed in the domestic context are similar from day to day and from group to group (ibid). For example, food preparation and consumption is a basic group maintenance task that is usually repeated during an occupation (assuming, of course, that food is available). Between sites, dumps should be structurally similar, and have the following general characteristics although the artifact classes involved might vary: hearth area dumps tend to be (1) relatively rich deposits, (2) associated with a feature, and (3) have contents that reflect many different activities internally (have a wide variety of artifact classes).

The formation process characteristic of hearth contexts (the gradual accumulation of refuse through repeated disposal events in the same place) is primarily responsible for the content characterics noted above. This form of refuse control may also occur in nondomestic contexts when conditions dictate (e.g., when different tasks requiring clean surfaces are performed in sequence or when the presence of several social groups requires each group to clean up its refuse from common use areas). If refuse control is a prevalent deposition mode at a site, all localized deposits should be similar in content, with any differences related only to the number of objects composing a concentration (sample size). In other words, the repeated disposal of debris from many tasks in few controlled locations, regardless of where the activities were actually performed, will result in a series of material concentrations over the site surface whose contents are homogeneous when all are compared (after accounting for sampling effects).

The second kind of deposit characteristic of hunter-gatherer settlements is the peripheral or extensive activity area. These have particular locational attributes according to both Binford and Yellen. As their name implies, peripheral areas are situated outside of and away from domestic hearth areas. Thus, they are "peripheral" relative to the core areas. Peripheral areas are typically the loci of specific activities that, although they may be repeated during an occupation, are usually less ubiquitous and have more spatial integrity than group maintenance tasks.

Formation processes in the peripheral zone also differ from those around the hearth. Deposits resulting from activities in peripheral areas often comprise waste materials left in place, since these are already out of the way of intensive human traffic during the occupation. Also, material accumulations in the peripheral zone are often the result of a single activity event (Yellen, op cit.). Thus, while tool deposition in these areas may not accurately reflect the "tool kit" used in task performance (for the reasons mentioned in Chapter 1: curation, loss and dropping rates, etc.), waste materials here may be directly related to the activity. Binford argues that these areas may have a circular pattern as opposed to fan-shaped hearth dumps (1983: 165-72), but this seems particularly contingent on context, as the shade-seeking !Kung illustrate.

In content, peripheral zones should be composed of restricted sets of artifact classes. More importantly, peripheral activity area deposits should be externally heterogeneous. This is because they have distinct contents when they result from different tasks and differ from general hearth deposits in terms of formation processes (Yellen 1977: 131). Thus, peripheral area deposits may inform most directly on the reason for site occupation, since they represent the "special", nonmundane aspects of site use.

This brief discussion based on actualistic observations has allowed some general characteristics of hunter-gatherer spatial organization to be defined. These include two different kinds of deposits generated at sites, two distinct processes of refuse accumulation, and some expectations for how these might be distributed in space. Armed with these new tools, variables relevant to modelling site formation at Pincevent and Le Flageolet can now be discussed.

VARIABLES FOR CHARACTERIZING SITE FORMATION PROCESSES

Based on ethnoarchaeological observations, the content of deposits resulting from various formation processes can be expected to vary in different ways. For example, specialized activities performed at the periphery of an occupation area will tend to result in archaeological deposits that are made up of waste products, tools broken in task completion, and lost items of continuing utility related to the activity. Each separate deposit will be as distinct from any other as are the tools, raw materials, and probable tool breakage rates involved in tasks performed there. On the other hand, core area dumps tend to comprise materials generated over longer periods of time, represent a variety of activities, and form by frequent and repeated events of refuse disposal. Each dump formed in this way will tend to be similar to others generated in the same way.

When the site surface is considered as a whole--the contents of all clusters or accumulations are compared--the relative importance of the first kind of deposit (activity areas) to the second (refuse dumps) will be reflected in the homogeneity of cluster content, defined as the similarity in content exhibited by all spatially discrete material accumulations over an occupation surface. If accumulations are homogeneous, then most were generated by dumping (the "Homogenizing Process"). If deposit content is heterogeneous then many accumulations relate to more specific kinds of activities, and contents therefore vary from one deposit to another. Thus, one dimension that can be used in modelling variation in site formation process is the homogeneity of artifact cluster content. The measurement of this variable will be discussed in Chapter 3.

A second dimension that will be involved in characterizing site formation concerns expected variation in the sizes of areas involved in different depositional processes. As we have seen, at least two ethnoarchaeologists suggest that core areas are essentially constant in amount of space devoted to the hearth and dumps. This consistency may be related to the number of people normally composing a family or task group in hunter-gatherer societies or may have an even more fundamental constraint in the size of the human body (Binford 1983). In contrast to domestic areas, peripheral activity areas can be quite varied in size, depending on the space required for each activity. Hide processing areas, for example, will vary in size depending on the number of hides, the animal species, and the number of people involved. Roasting a porcupine, requiring pit construction, takes a different amount of space than processing mongongo nuts (Yellen, *ibid*). In short, different site formation processes may result in accumulations of different sizes. This problem was recognized by Kintigh and Ammerman (1982), and, in part, motivated their development of the *k*-means approach. More recent work has illustrated how this problem might be treated analytically (Simek and Larick 1983).

Thus, when distributions involving a single site formation process are analyzed, techniques must be able to identify the relevant *level of spatial refinement*, defined as: *the size of spatial concentrations characterizing a pattern.* It follows that sites generated by more than one formation process require analysis in terms of all the appropriate levels of spatial refinement. This presents a technical problem which, it will be shown in the next chapter, has been solved. It also requires the assessment of homogeneity relative to all levels of spatial refinement that may be relevant to the different formation processes. For example, when a !Kung San occupation site is considered at a low level of spatial refinement, so that the accumulations being assessed for homogeneity take up a large part of the site surface, content among concentrations is homogeneous and reflects recurring family-size domestic areas along with nearby peripheral areas. On a higher level of refinement, !Kung sites are more heterogeneous in terms of accumulation content. This is because smaller units of space compose the pattern, and dispersed activity areas are defined as separate entities. Clearly, homogeneity must be considered on a variety of levels of spatial refinement since depositional processes vary in this dimension.

SIMPLE MODELS FOR VARIATION IN DEPOSITIONAL PROCESSES

Based on the foregoing discussion, the dimensions "homogeneity" and "spatial refinement" are employed to produce models for archaeological site formation to be used in the case studies that follow. Two-dimensional representations are employed. The vertical axis represents homogeneity in content among concentrations (measuring this axis for an actual distribution is a complex problem addressed in the following chapter on analytic techniques). The horizontal axis represents changes in the refinement of space: "low" refinement refers to large spatial units (more inclusive of space) and "high" refinement means small and usually more numerous units. Refinement is thus synonomous with "resolution": higher resolution reflects more detail, and the distribution is more refined.

It should be stressed that the models of process presented here are only a few of many potential relations between the variables that could be imagined. This is because, as was stated earlier, models were developed in part to be relevant to the particular contexts and analyses of Pincevent and Le Flageolet. The primary purpose for these models is as heuristic devices that illustrate specific situations. Given the nature of the variables included in model formation, it is highly likely that interrelations are potentially continuous and that the models presented here may require modification or replacement before they can be assessed in other archaeological contexts. However, I would suggest that the variables themselves might be useful in other situations.

In modelling possible relations between accumulation homogeneity and level of spatial refinement, four discrete situations will be considered. The first represents a reformulation of the traditional activity model, wherein all material accumulations over a site surface represent distinct activity deposits. In such a situation, a single depositional process is responsible for all material concentrations in space: the disposal of objects produced or used in a spatially isolated task at or near the locus of generation. This kind of deposition might occur in places where occupation involves the rapid performance of specific tasks with little or no group maintenance concurrent; for example, a kill site might exhibit such specific patterning (Wheat 1973, Binford 1978b). Figure 1 illustrates the expected relation between homogeneity among accumulations and spatial refinement in such a situation: homogeneity is low at all scales. The form of this relationship reflects the great variation in content among accumulations over the site surface regardless of such factors as sample size or the size of the spatial unit defined.

A second model reflects a situation where group maintenance is an important factor in spatial process but where specialized activity performance is also substantial. Here, deposits are homogeneous at a low level of spatial refinement, reflecting the number of social groups utilizing a separate "domestic core area". At higher levels of refinement, spatial units are more heterogeneous, reflecting the variation introduced by activity areas that were not defined as separate spatial units at lower levels. Figure 2 illustrates the expected relation between the model variables when accumulations seen at a low level of refinement are homogeneous while small spatial units are heterogeneous. This form of model might be especially appropriate for identifying sites where several similar social units, each occupying a discrete domestic area, were present during occupation. This model characterizes those !Kung San settlements described as habitation sites by Yellen (loc cit.).

Figure 3 illustrates a situation opposite to that modelled in Figure 2. Here, large spatial units are heterogeneous while small units are homogeneous. Such a relation is difficult to envision in behavioural terms and would require: (1) that large areas of a site were reserved for specialized activities, (2) material classes involved in each activity were the same with continuous variation in class frequencies between areas, and (3) the refuse generated within an activity area was disposed of in a controlled fashion (dumped in the terminology employed here) thereby homogenizing small accumulations. This kind of deposition was not encountered in the course of research at Pincevent or Flageolet but is presented here for comparative purposes.

The fourth and final formation model is presented in Figure 4. Here, both large and small accumulations are homogeneous in content, representing the opposite of the situation discussed in Model 1. Such a form might result in occupations where systematic dumping is practiced. Refuse from a variety of activities, perhaps discretely organized in space, is collected and deposited into a few locations that are undifferentiated as to the context of material generation. Small spatial accumulations represent "pieces" or little versions of larger ones. A key here is the single formation process operating; this process results in the mixture of materials resulting from a variety of activities.

The four models illustrated in the preceeding graphs are intended as guidelines for investigating site formation processes on an analytic level. They represent four extreme conditions and can therefore serve as bases for evaluating observed relations between accumulation homgeneity and level of spatial refinement. In the studies of archaeological distributions presented in Chapters 4, 7 and 8, these models allow the assessement of formation processes in the sites and some measure of comparison of differing structural properties among the archaeological cases. Before turning to the case studies, however, Chapter 3 will discuss the techniques employed to operationalize the concerns expressed in this and the preceeding chapter.

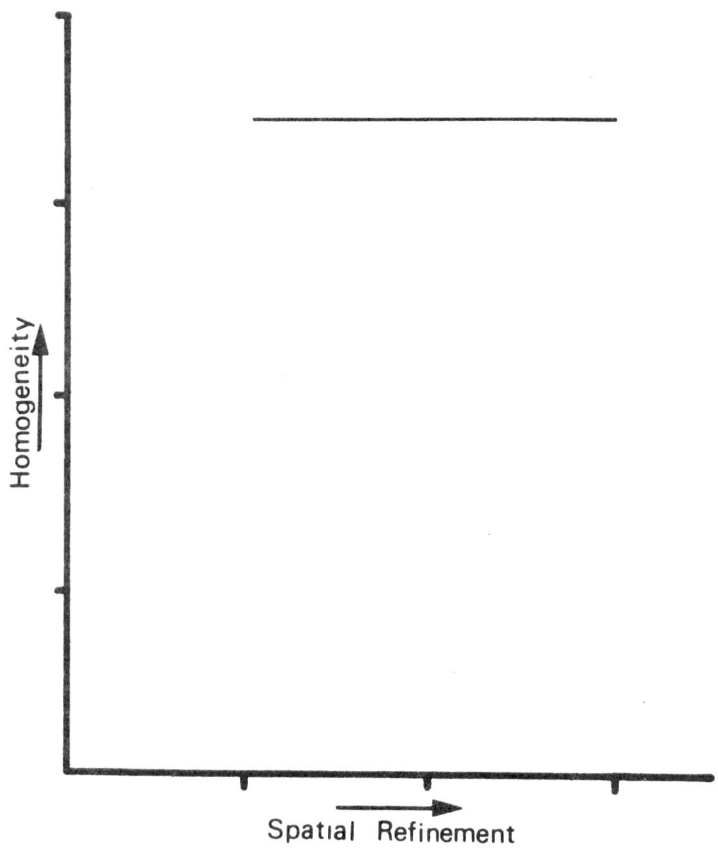

Figure 1. Expected Relation between Homogeneity and Spatial Refinement: Model 1.

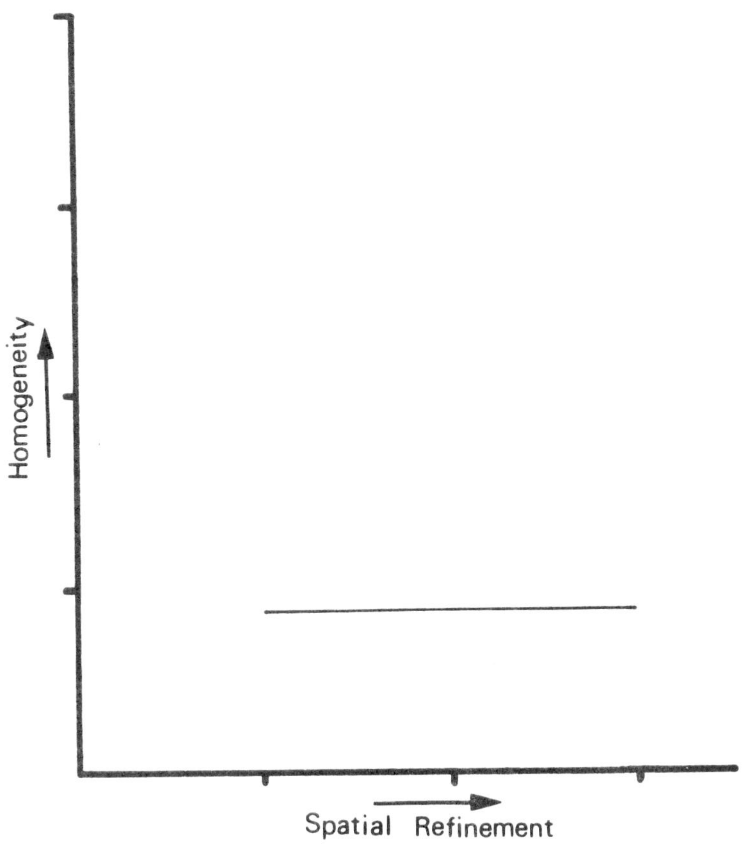

Figure 2. Expected Relation between Homogeneity and Spatial Refinement: Model 2.

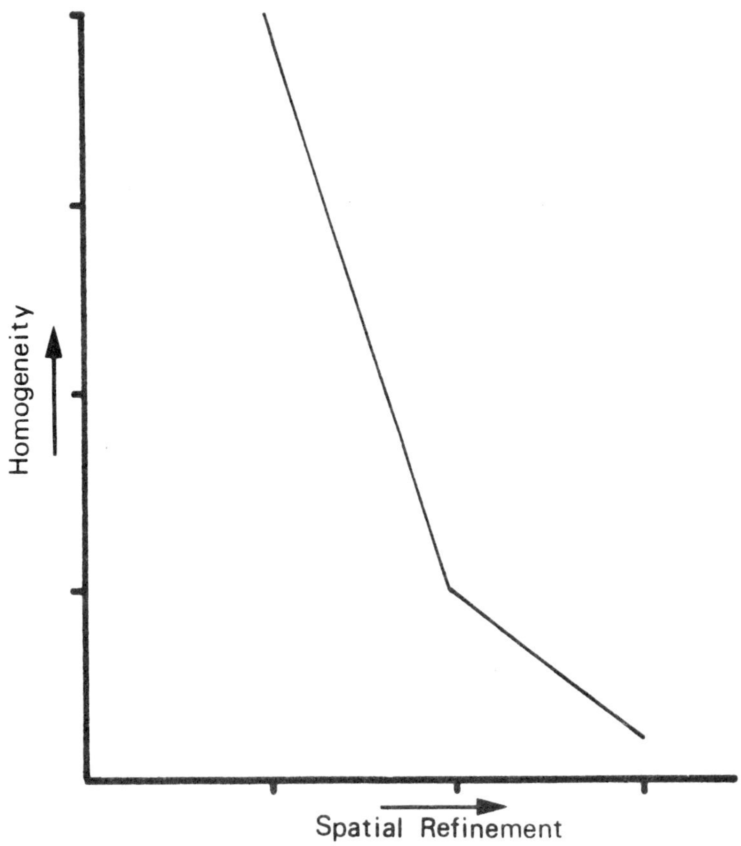

Figure 3. Expected Relation between Homogeneity and Spatial Refinement: Model 3.

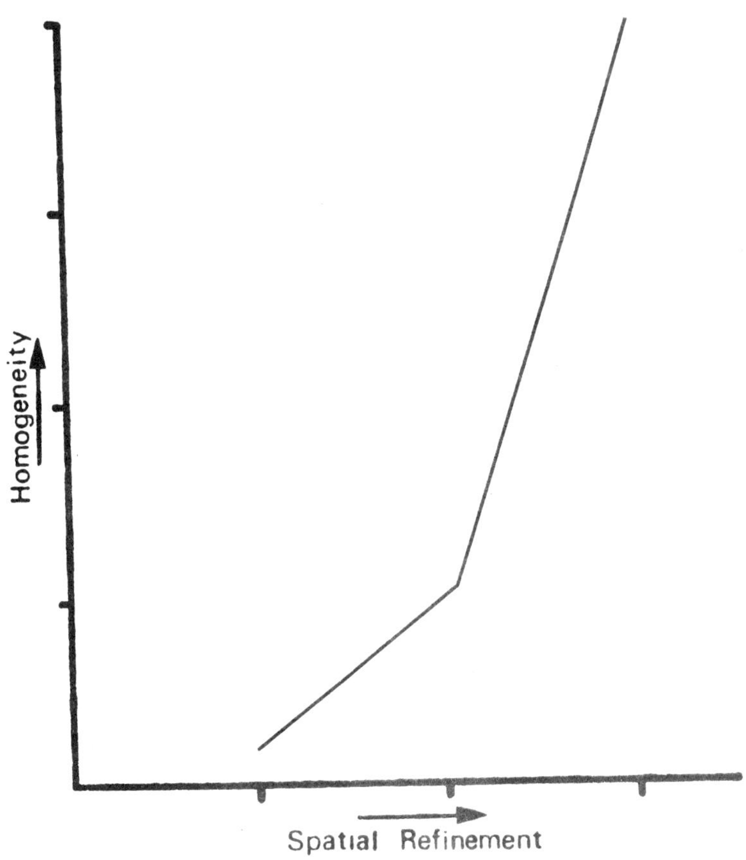

Figure 4. Expected Relation between Homogeneity and Spatial Refinement: Model 4.

CHAPTER 3

METHODS OF ANALYSIS

INTRODUCTION

In order to evaluate the models presented in the preceeding chapter, and to address the ultimate goal of understanding paleolithic site use at Le Flageolet and Pincevent, analytic methods that are consonant with the modelled variable interactions are required. In short, these methods must allow the complex variables "homogeneity" and "spatial refinement" to be measured. Clearly, methods employed toward this end will be varied, since each variable relates to data of somewhat different form. For example, spatial refinement must involve some characterization of location (position) over the site surface, while homogeneity need not. As a first step, a technique for recognizing spatial patterns within distributions is needed (to define units that can be examined as "accumulations" at a single spatial scale). A variety of methods proporting to do this are available. However, not all pattern recognition techniques employ the critical variable location; those that do not are of little use in the present research context. Techniques previously used to address the problem of pattern recognition in archaeology will be discussed below, and the choice of method applied here will be justified.

In addition to pattern recognition on a single spatial scale, analytic methods designed to assess the formation processes discussed in Chapter 2 must be able to identify patterning on a variety of levels so that these can be compared. It will be remembered that expectations are for multiple site formation processes at a single location, each of which may operate or be expressed on a different locational or size scale. This is of considerable importance given the data sets analyzed here; Le Flageolet in particular has already defeated some methods because of its contextual complexity (see Chapter 1). As will be discussed shortly, a method that satisfies these requirements is available; pattern recognition can be pursued at a variety of spatial scales without losing information at any one scale.

The problem of identifying and characterizing homogeneity among spatial accumulations requires a different approach than basic pattern recogniton, since data are of frequency rather than locational form. To examine homogeneity, pattern recognition methods must result in meaningful units within and between which variation in content can be assessed. In short, counting units with spatial relevance are required that can be compared both in terms of abundance (e.g., the absolute numbers of items) and in terms of variety (i.e., referring to different classes of material). Finally, the last two charactersitics must be combined to form a homogeneity measure that can be examined over space. These requirements are, in fact, less demanding than they might appear when considered abstractly. Methods for carrying out all of these analytic phases are applied here, and most are rather simple in statistical terms.

Before discussing the actual system used in the studies of Pincevent and Le Flageolet, however, a brief discussion of conventional spatial analysis in archaeology (with special reference to techniques used to carry out both pattern recognition and association studies) is appropriate. The methods ultimately selected for use were chosen because they (1) address the perceived problem at hand in ways other techniques do not, and (2) seem more consistent with modern concerns and expectations for the nature of the archaeological record than do previously-applied methods. For this reason, other techniques and the reasons for their elimination from use in these analyses will be examined. But first, a few words concerning the nature of archaeological spatial data, and in particular the data collected in recent excavations of paleolithic sites, are presented to set the background for the discussions that follow.

THE ANALYTIC CHALLENGE: ALL THAT DATA

Archaeologists undertaking a spatial analysis of their excavation data are frequently confounded by several analytic problems. Modern data recovery techniques are one source of these problems. Recent excavations, particularly those carried out in paleolithic sites, stress the recording of all potential information. Often, this includes the taking of three dimensional coordinates for much or all of the artifactual material unearthed in a site. These spatial data are usually transformed into point distribution maps of sometimes daunting complexity, especially when thousands of artifacts are plotted. As a rule, preliminary data reduction is carried out, often comprising the division of very dense maps into series of less-complex class distributions. These classes might reflect functional classification of tools, animal skeletal parts, or even aspects of artifact attributes. For the spatial analyst, the site is thus represented by not one, but many point plots. The task is to examine these for overall patterning in the distributions, patterns within individual class plots, and associations in space between artifact classes.

The complexity of the analytic problem obviously precludes the use of a single method. Because a series of different questions is being asked of the data, various systems for analysis are required, each designed to address part of the question. At the outset, point pattern maps must be examined for patterning in the distributions. This aspect of spatial analysis can be termed "pattern recognition" and may have several levels of relevance within a single study. For example, pattern recognition is required for the distribution of all recovered materials (without defining classes) to examine the organization of the site as a whole. Such a "global" organization may or may not be present, but if it is, it might condition patterns within specific artifact classes. To identify such a relationship, pattern recognition must also be carried out within class distributions.

Once patterns have been identified, the relationships among different artifact classes over space can be examined in terms of the extant patterning. Of concern here is the association among artifact classes over space. This analytic phase can be termed the "pattern association" stage of spatial analysis, since one seeks to compare spatial patterns from different classes in search of regularities.

PATTERN RECOGNITION: PREVIOUS APPROACHES

Informal Pattern Recognition

Two forms of pattern recognition have traditionally been employed in archaeological research. The oldest and perhaps most prevalent technique involves the informal visual examination of distribution maps in search of patterning. The visual approach has many benefits when properly carried out, not the least of which is plausibility. At its best, visual inspection may be the most satisfying method available. Given sufficient time, all possible permutations and combinations of the data can be studied, patterns can be found and compared, contextual elements can be integrated with observed patterns, and areas within a site that seem to have meaning in light of the excavator's expertise can be isolated. Indeed, the visual approach allows the most complex relationships between empirical facts and the archaeologist's knowledge to be established of all spatial analytic techniques. This is because the method is nonmechanistic and leaves room for creativity in analysis and interpretation. However, with the kinds of dense distribution maps generated by today's excavation techniques, the inspection approach is prone to "data overload" so that complete analysis can be too complicated using intuition alone.

Visual inspection is, almost by definition, very time consuming. More importantly, adequate evaluation of patterning (and subsequent interpretion) requires analytic input from a veritable army of specialists. Specialized studies are necessary since the derivation of informal inferences depends on a complete understanding of all factors that might influence a given distribution. This is not to say that archaeologists using quantitative methods for pattern recognition do not require specialized information; they simply do not need that input for pattern recognition as is the case in the visual method. In the informal approach, "pattern" and "meaning" are synonomous. Patterns are identified only when they are "reasonable" in terms of what the analyst already knows. The analyst therefore requires as complete a knowledge of the site as possible so that unreasonable patterns can be ignored. Visual inspection, despite its time-consuming nature, has produced remarkable synthetic interpretations at sites like Pincevent (Leroi-Gourhan and Brézillon 1966, 1972) and Le Flageolet (Delpech and Rigaud 1974; Rigaud 1980).

Despite potentially satisfying results, the visual approach defies formal replication, and success must be judged based on the eminence and capability of the analyst. Comparison among interpretations offered by different archaeologists is virtually impossible because the criteria for pattern definition are rarely equivalent. What constitutes a pattern, what is included in a cluster, and how associations among artifact classes are defined; are all subject to variation in definition and application. We are left with a series of "plausible scenarios" that are difficult (if not impossible) to relate in a systematic, synthetic fashion.

Where the visual approach is, perhaps, weakest is in its ability to detect multiple or hierarchically-related patterns. Given the complex natural and cultural factors that might have operated to produce spatial patterning at an archaeological site, it is somewhat naive to expect a single pattern to account for all the extant spatial variability. It is more likely that simple patterns (those that are manifested over large areas of the site surface) contain finer-scale patterning. Visual inspection is more than able to detect large-scale patterns, but these are often the only ones seen. Leroi-Gourhan and Brézillon's work at Pincevent (ibid) and Wilmsen's studies of the Lindenmeier site (1974, 1978) represent examples of this problem. In both cases, patterns in the spatial distribution of objects were detected and interpreted as reflecting "household" units. Smaller-scale patterns within the households went unrecognized, probably because the human eye is "held" by obvious patterning. Collingwood (1946) relates this problem to the Gestalt effect defined by psychologists. It seems likely that this is a general problem: the human eye is blinded to patterns within patterns once the more obvious forms are seen because it becomes fixed on the more evident configuration.

Formal Pattern Recognition

As an alternative to visual inspection, recent spatial analyses of archaeological data have relied on quantitative techniques, for the most part borrowed from other fields of study. These methods have the advantages of formal definition of terms and relationships, replicability, and the ability to detect patterns that might not be visible to the human eye. Since most of the quantitative methods employed in archaeological spatial analysis are computer-assisted (or potentially so), they are able to handle large numbers of data cases in much less time than required for visual analysis.

Quantitative approaches applied to archaeological pattern recognition can be divided into two main types: those which employ data reduced to quadrat counts and those which act on point data. An important characteristic of most methods, of both kinds, is that they were borrowed from disciplines other than archaeology that have longer traditions of spatial analysis. Many techniques were developed for applications to ecological problems, while others were originally designed to address questions posed by geographers, geologists, and economists. Hodder and Orton (1976) provide a general survey of many of these methods, so a detailed digression is unnecessary here.

One problem inherent in the use of borrowed technology concerns the relevance of assumptions underlying the procedures. For many methods developed in plant ecology (e.g., Cormack and Ord 1979; Kershaw 1964; Pielou 1977), the underlying goal is to detect random and non-random distributions of plants within a given area of land. This goal is based on known processes of seed dispersal and the assumption that, all things being equal, these processes will produce random spatial distributions. Non-random patterns of plant locations, thus, indicate that all things are <u>not</u> equal, and the ecological constraints on dispersal can be investigated.

Archaeologists have used these kinds of methods without critically evaluating the purpose of the tests. Dacey's (1973) application of the variance/mean ratio test (based on a Poisson model for random frequency distributions) to the problem of archaeological distributions is an example of this kind of borrowing. The goal of Dacey's study was to detect the presence of non-random artifact distributions over a site surface. He never adequately explains why a random model is appropriate in archaeological situations, and we already know that such distributions are the results of non-random formation processes (human behaviour, natural erosion, etc.). Rejection of the null hypothesis in this test (indicating that the distribution under study is not random) tells us no more than we already know; acceptance of the random hypothesis is in opposition to the facts. Moreover, as Hietala and Stevens note (1977: 539), these tests do not consider possible spatial association between different maps (e.g., of different artifact classes) that may individually show statistically random distributions. In short, the utility of tests for random distributions in archaeological spatial analysis must be questioned.

Quadrat Techniques. Quadrat-based methods have been proposed that search for clustering within distributions without reference to the randomness of the global configuration. Most of these assess clustering of frequencies at a variety of different block sizes based on an original grid of cells and test for changing variance among cells between block sizes (e.g., Whallon 1972, 1973). These methods have been criticized for a variety of shortcomings (Doran and Hodson 1975; Hodder and Orton 1976; Hietala and Stevens 1977). The biggest problems lie in the methods' lack of sensitivity to clusters of varying sizes and shapes over a single distribution and in their dependence on grid units of standard shape. Because of these problems, other factors (for example, the stipulated size and original position of the grid cells and the locations of grid boundaries) affect results of the tests more than is desirable.

Several quadrat methods have recently been proposed that were designed specifically with archaeological problems in mind. However, the basic machinery for these techniques derives from other research contexts. Still, interesting results have been attained. Among such treatments are "Local Density Analysis" (Johnson 1976) and the probability-based methods suggested by Hietala and Stevens (1977). Of these, Local Density Analysis is the least satisfying, particularly when associations among artifact classes are assessed. The technique employs Principal Components Analysis to detect cooccurring artifact classes over a site surface, thereby assuming that a linear model is appropriate for these kinds of relations. Such an assumption is probably unwarranted and must in any case be demonstrated before application of linear models (Speth and Johnson 1976).

Hietala and Stevens' method has a stronger base statistically. However, the requirements of statistical tests for association often result in the aggregation of small grid units into larger cells when sample sizes for certain artifact classes are small. In other words, when associations among classes are being assessed, all information must be scaled to the least informative level because cell counts for all classes must be aggregated until expected counts are sufficient for the rarest class.

Point Pattern Techniques. Most techniques previously applied to point pattern data in archaeology involve some form of nearest neighbor measure. These methods retain some of the information contained in exact artifact provenience lost by reduction to quadrat counts. However, many early applications of nearest neighbor techniques suffered from the statistical effects of arbitrary excavation boundaries that severely impacted analytic results (cf. Whallon 1974). More recent applications of nearest neighbor statistics include correction factors designed to overcome boundary effects (e.g., Pinder, et al. 1979). Yet, a basic flaw remains. Nearest neighbor analysis results in a single statistic that can be used to characterize the point distribution under study as "clustered" (aggregated), "random", or "regular" (uniformly spaced). Regardless of the state to which a distribution is assigned, its spatial aspect is thereby lost. Thus, even though a distribution might be identified as "clustered", the method provides no means for locating clusters in space.

This last problem is an important one, and it characterizes most of the borrowed techniques that have been employed by archaeologists. Again, this is due to the nature of the problems for which the methods were originally designed. In plant ecology, results indicating that a distribution is random, clustered, or uniform have clear and distinct meanings. Spatial location is not important, since the tests are carried out to delimit potential ranges of explanation. But in archaeological spatial analysis, we must know where clusters occur before we can begin to infer their meaning. This is true at both an intrasite and a regional scale of analysis. Our explanatory frameworks for spatial distributions involve location. This is because we do not know the causal processes operating to produce any spatial configuration. Plant ecologists do. Archaeologists must infer process once the nature of a distribution can be related to other aspects of the archaeological record. To accomplish this, we must know how patterning is manifested, i.e., located, over the space being analyzed. Only thus can other characteristics of the record at the same places be related to discovered spatial patterns. Quantitative techniques that eliminate location from results therefore deprive us of our inferential base.

Useful Techniques. It would seem, then, that most of the quantitative treatments previously applied to pattern recognition in archaeology do not provide the kinds of information necessary and relevant to the problem. These techniques are often insensitive to variability in the size and shape of material concentrations, reduce excavation data to the point where critical information is lost, and eliminate location as a dimension of interest in analyzing spatial distributions. Such problems clearly make the use of these kinds of methods inappropriate for archaeological concerns. To adequately address archaeological interests, relevant pattern recognition techniques must satisfy at least four critical criteria:

1) A certain amount of data reduction must occur, since typical, modern excavations produce overwhelmingly dense distribution maps. Such data reduction must involve the recognition and evaluation of patterns in spatial distributions.

2) Pattern recognition techniques should be quantitative for two reasons. First, the massive amounts of data produced in today's excavations can only be handled in quantitative fashion. Second, solutions that are formal and replicable can be judged objectively and tested empirically.

3) Locational information obtained in excavations should not be ignored in analysis. Results should be comparable to characteristics of site locale, features, topography, etc. Thus, in addition to the data reduction called for in (1), one might add that too much reduction is undesirable.

4) A useful analytic technique will allow identified patterning to be mapped in terms of the size and shape of constituent units (e.g., material concentrations). These units will also allof comparisons among different artifact classes and between sites. Without such an ability, the identification of any associations among classes over a site surface is impossible since the scale and position of association cannot be defined.

Of all the spatial analytic methods so far proposed for archaeological pattern recognition, only one satisfies these criteria while avoiding the time required for informal visual inspection. This method, based on a <u>k</u>-means clustering algorithm, was developed and proposed by Keith Kintigh and Albert Ammerman in a recent <u>American Antiquity</u> article (1982). It is the technique that will be used in the spatial analyses composing this volume.

PATTERN RECOGNITION: THE HEURISTIC APPROACH

The basic goal of Kintigh and Ammerman's method is to build,

> a number of intuitively derived heuristics into an analytically useful computer program...We wish to construct a program that will examine spatial data and derive concise higher level information concerning their distribution [1982: 34].

The heuristic nature of the method combines the descriptive rigour of quantification with the elegance and clarity of intuitive pattern interpretation. Indeed, it is the interplay between an analyst's knowledge about a site (e.g., of contextual features) and the results of mathematical characterizations of the distribution under study that gives the approach its heuristic quality. By "heuristic" then, Kintigh and Ammerman refer to the use of quantitative techniques as means for describing data so that the archaeologist can occupy his or her time with understanding--interpretation--of the record. They are allowing for the fact that the human mind can often sort out important patterns from a series of possible ones better than can a machine.

K-means Cluster Analysis

As the basic pattern recognition tool, a *k*-means clustering algorithm is used (Hodson 1971; Johnson and Johnson 1975). According to Hodson (*ibid*: 31):

> the *k*-means method of cluster analysis seems the most promising approach to follow for large quantities of data, provided that relatively few variables are concerned and a relatively small number of groups is suspected.

More specific advantages of the *k*-means method for archaeological analysis will be discussed below. The general characteristic of the technique--an ability to handle large data sets inexpensively--is an important one in archaeological spatial analysis.

In the *k*-means program implementation developed by Keith Kintigh, a series of supplementary statistics are produced that are of great utility for spatial analysis. These help to overcome some of the early problems with the method noted by Hodson, including a tendency to find "local" rather than "global" solutions (Hodson 1971: 32). These statistics provide locational information and characterize the size and content of spatial concentrations within the basic reduced pattern description resulting from *k*-means analysis. Other statistics allow the production of display maps for results. These maps can be compared among various partitions of the distribution, for example, by artifact class. A complete discussion of the *k*-means method and program is presented in Kintigh and Ammerman, but the relevant aspects of the technique will be discussed here as a background for the applications which follow.

While *k*-means cluster analysis is not entirely new to archaeology, it has not previously been applied to problems of spatial analysis. This is somewhat surprising considering the advantages described by Hodson. The method has been used primarily in artifact classification studies (Hodson 1970, 1971) and as a technique for chronological seriation of ceramics (Johnson and Johnson 1975). In previous applications, emphasis has been on the data handling capacity of the algorithm. In other words, *k*-means has been used when data sets are so large as to preclude the economical use of other clustering algorithms. Because the program acts directly on the input variables there is no need to calculate and store similarity matrices to measure relations among the data cases. A more direct measurement among cases is made--in this case the linear distance between objects in geometric space--so that massive matrix production and manipulation is unnecessary. This is precisely why *k*-means can handle very large numbers of points. That *k*-means reduces computation costs is one of its more mundane attractions. More interesting from an archaeological perspective, the method can characterize spatial patterns with reference to the actual space involved and to the relations among data points within that space. No intermediate data transformations are required that involve unwarranted assumptions about data form or distribution characteristics.

The *k*-means algorithm has the advantage of simplicity. Moreover, the technique has three characteristics that differentiate it from other clustering methods applied in archaeology, most notably single- and complete-link cluster algorithms (cf. Hodson 1970; Sneath and Sokol 1973). The method is divisive rather than agglomorative. This means that the distribution is initially considered as a whole and divided into smaller and smaller clusters; agglomorative algorithms begin with all individual points and build clusters sequentially until a single cluster results. Because of this, boundary effects are overcome with *k*-means since the initial shape of the distribution provides the analytic space. *K*-means analysis is also overlapping in nature. An individual case assigned to one cluster at a certain point in the analysis (according to the proximity of that case to the center of the cluster) can move to another cluster later in the run if the distribution is characterized better by that reassignment. Finally, the *k*-means method is nonhierarchical, since cluster configurations produced at one level do not determine configurations at other levels. In other words, clusters defined late in the analytic sequence (when more clusters are being formed) are not nested in or dependent on earlier solutions. The problem of "chaining" that affects so many other clustering techniques (Sneath and Sokol 1973) is thereby avoided. At the same time, the dendogram method of illustrating clustering may not be a relevant way to display *k*-means results.

These characteristics of the *k*-means algorithm overcome many of the problems inherent in previous clustering methods used in archaeology. Its divisive nature overcomes boundary effects. By beginning with the distribution itself as the initial analytic space, *k*-means is useful for analyzing a variety of spatial data, including linear distributions such as might characterize riverine settlement systems. (This utility is demonstrated in Kintigh and Ammerman 1982). Because *k*-means can handle large data sets, point pattern data can be analyzed in their pure form without initial data reduction and resultant loss of information. Allowing overlap between clustering levels insures that points are assigned membership in the closest cluster at each analytic level. When this aspect is combined with the nonhierarchical character of *k*-means, a sequential process of analysis results whereby a series of solutions is produced, any one of which can be considered independent of those before or after it. This is an important advantage when several scales of patterning are anticipated within a single distribution.

How K-means Works

The *k*-means algorithm, in its current implementation, has several parts. The first is called the "split" phase, and it is here that clusters are actually formed. At the outset, the center of the whole distribution is located by computing the mean value for each coordinate. The point thus defined is called a *centroid* since it represents the center of a *cluster* (here the entire distribution or the *one cluster level*). Next, the furthest point from the centroid is found by calculating the distances between it and all data points. This farthest point becomes the "seed" for a second centroid, i.e., at the two cluster level. The distances between all data points and the seed are computed, and those points closer to the seed than to the original centroid are assigned to a new cluster including the seed (distances to the original centroid are already known). The centroids of both new clusters are computed by averaging *x* and *y* coordinate values for member points, and another pass is made through the data measuring distances and reassigning data points to their nearest centroid. This process is carried out until no more reassignments are made, i.e., a *stable solution* is reached. By this time, the centroids have "migrated" until they locate the centers of the two stable clusters. The result is a *two cluster solution* for the distribution under examination.

After the distribution has been characterized at a given level, the resulting clusters are "lumped" back to the previous solution level by combining the two clusters with the closest centroids. This lumped configuration is compared to the previous solution of the same number of clusters using the SSE statistics computed for each level (see below). If the first solution has a lower SSE than the lumped solution, it is retained and lumped again. If, however, the lumped solution is "better" than the original (has a lower SSE), then _it_ is retained, and the search for one more cluster proceeds by splitting. The process of cluster formation and evaluation composes a program loop. In order to achieve the "best" solution for a clustering level, lumping must proceed all the way back to one cluster. When this occurs, it means that all retained solutions are the best ones possible. The solution obtained at the highest level during the pass (two clusters in our example) is then used as a basis for a splitting phase seeking the next highest solution.

The _k_-means program proceeds in this manner from one to two to a user-specified terminal number of clusters (hence "_k_ means"). At each clustering level, the best solution is found before going on to the next level. The criterion for successful clustering at each solution level is defined using the _SSE_. This statistic is calculated as the sum of the squared distances between all data points and their cluster centroids. The SSE, therefore, represents a sum of squares error measure for each solution. The _k_-means procedure minimizes this measure during the splitting and lumping phases at each level. By finding the lowest value for this statistic, the most appropriate characterization of the distribution under study is achieved.

Defining Relevant Patterns

It is unlikely that relevant patterning for a given distribution exists at all the solution levels defined during a _k_-means run. For example, even though the _k_-means program will produce best descriptions of a distribution at from one- to fifteen-cluster configurations, relevant patterns might occur only at the two and eight cluster solution levels. Again using the SSE statistic as a basis, the entire sequence of solutions produced during a run can be assessed in relative terms to find the _optimal solutions_--those that actually conform to patterns in the data. Here is how this is accomplished.

At the end of a _k_-means run, a series of SSE statistics is produced, each one representing the minimum sum of squares at at given cluster level. Some of these (the ones that represent actual pattern levels) are "more minimal" than others. To determine which are the optimal solutions, the SSE for each level is expressed as a percentage of the total SSE in the distribution. This value is then transformed by log10 for convenient display. The transformed statistic (the log(%SSE)) has a scale with an upper limit of two (a value corresponding to the entire distribution considered as a single cluster) and decreasing values from this maximum through the solution sequence. The log(%SSE) can be expected to decrease naturally as more and more clusters are formed, since the distances between data points and centroids decrease with more centroids. However, when substantial improvement in the description of patterning occurs from one solution to the next, an inflection, or drop, in the log(%SSE) also is apparant. By plotting the log(%SSE) against the sequence of clustering levels for all solutions in a run, such inflections are displayed and identify the optimal solutions for a given distribution.

Figure 5 illustrates the log(%SSE) plot produced by a k-means analysis of burin distributions at the Upper Paleolithic site at Pincevent (Leroi-Gourhan and Brézillon 1972). Inflections in the plot are evident at the three, six, and fourteen cluster levels. From this, three optimal solutions for the burin distribution can be defined. It should be pointed out that the relative strength of inflections cannot be taken as indicative of the strength of the patterns themselves. Even though the three cluster solution shows the most severe drop in log(%SSE), this does not mean that the three cluster solution is somehow the "best". As was mentioned above, there is more correspondance between the total SSE and a solution level SSE when fewer clusters are considered. This is because the distances between centroids and data points are by definition greater with fewer centroids. When more clusters are formed, there is less variation available from level to level, and inflections appear to have less magnitude later in the clustering sequence.

Just the same, this feature of the k-means program is very important since it provides a means for recognizing small scale patterns within larger, perhaps visually more obvious, ones. Given stated expectations for intrasite spatial analysis (Chapter 2), the k-means procedure satisfies another of the basic requirements defined for a useful spatial analysis technique.

Several other statistics are generated during each k-means run that provide important information concerning the distribution under study. Cluster centroid coordinates locate the reduced data descriptions in space, thereby retaining the locational information contained in the initial point distribution. As has been discussed, these centroids are the average x and y coordinate values for cluster members. Since every point is assigned to a cluster at every level, the total number of member points can be counted for each cluster--the point frequency. This value gives an indication of cluster content. As a measure of cluster size and dispersal, an RMS statistic is

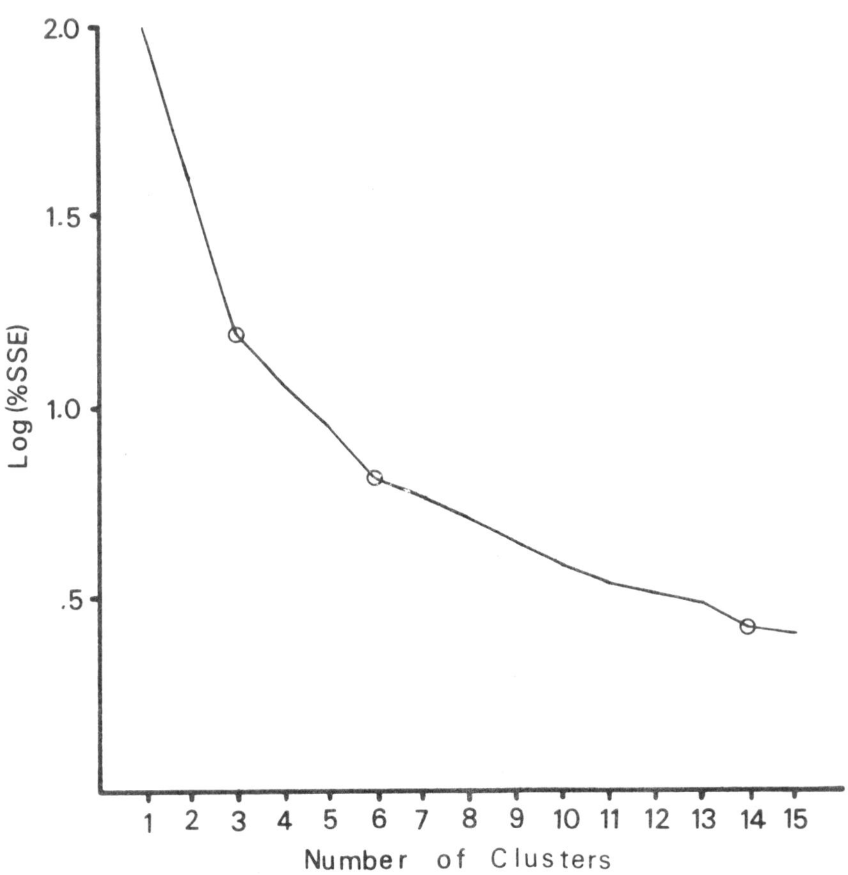

Figure 5. An Example log(%SSE) Plot Produced by Analysis of Burins from Pincevent Section 36 (Leroi-Gourhan and Brézillon 1972).

computed for each cluster at every level. The RMS can be considered a kind of spatial standard deviation of points assigned to a cluster centroid and is expressed as a radius within which 66% of all cluster members occur. A small RMS does not necessarily represent a small cluster as much as a compact concentration of many points. On the other hand, a large RMS radius around a centroid usually reflects a dispersed and often sparse scatter of points. These characteristics will be clear when maps of _k_-means optimal solutions are presented.

As a simple display technique, cluster centroids at a given solution level can be mapped on the site surface to locate the clusters. Then, the RMS values can be used to draw a circle around the centroids to show cluster sizes (actually relative local abundance of points and their dispersal pattern around the mean). Figure 8 in Chapter 4 shows such a map. By mapping all the solutions identified as optimal by the log(%SSE) plot for a _k_-means run, the multiple patterns characterizing a distribution can be displayed and examined.

In sum, the _k_-means procedure fulfills the requirements outlined above for a useful pattern recognition technique. In doing so, it overcomes many of the critical problems that have plagued other quantitative techniques applied in archaeology. When _k_-means is used, data reduction from massive point maps is accomplished with minimal information loss. The method has the advantage of being formal (thus results can be evaluated and replicated) and can handle large data sets quickly and economically. The _k_-means algorithm is simple and comprehensible. The descriptions provided by the technique retain locational information, give indications of cluster sizes, allow cluster content to be defined, and are easily displayed. The last advantage is very important because different solution maps can be overlaid (as will be seen) to allow direct visual assessment of pattern association and to allow the relating of artifact classes to contextual elements present on a site surface.

PATTERN ASSOCIATION: PREVIOUS APPROACHES

Informal Pattern Matching

Previous approaches to the search for pattern association in archaeological distributions can also be divided into informal and formal methods. Informal inspection usually does not make a distinction between pattern search and association studies, because they are most often carried out simultaneously. A now-classic example of this technique is J. B. Wheat's analysis of the Olsen-Chubbuck site (1972). Examining the contents of material concentrations observed during the excavation of Olsen-Chubbuck, Wheat was able to define several sets of bison body parts, bones from which were localized in space. Sets of stone tool classes were identified that associated with the bone sets. Wheat suggests that these spatial associations represent refuse from particular activities within the bison butchering sequence carried out at the site by paleoindian hunters. The informal approach to association study has both the strengths and weaknesses of informal pattern recognition.

Formal Pattern Matching

Many of the quantitative methods applied to archaeological distributions also address the problems of pattern recognition and association simultaneously. Several of these techniques are, in fact, more concerned with association than with pattern recognition, and it is for this reason that such important information as "location" is ignored. Most of these methods are probability based; they rely on various model distributions with known probability profiles. For the most part, the search for associations among artifact classes is carried out using the assemblage as the unit of analysis. Again, this results in location being eliminated from the equation. The assemblage emphasis has two main effects. First, even when spatial information is used initially in the analysis, results are usually in the form of a single statistic expressing the degree of spatial overlap between two (or more) distributions; spatial relevance is lost. Second, because many of the techniques look for overall associations among classes--using the assemblage as the analytic unit--the results are scaled to a global level. In other words, associations are identified only if they occur across the entire site surface. The possibility of smaller-scale or localized association is ignored. Such methods, thus, assume a monothetic profile for tool use and deposition (Clarke 1968). If, as is more likely, stone tools of given form were used in more than one task (cf. Keeley 1981), then the spatial scale of interest is much smaller than the entire surface under study.

An example of this kind of problem is Binford and Binford's study of Mousterian assemblage variability in which their "functional hypothesis" was first put forth (1966). They attempt to define "tool kits" among a series of Mousterian levels using Factor Analysis. By beginning with the assemblage as their unit for analysis, the Binfords eliminate spatial information *a priori* from their investigations. Although they might argue that their study was not "spatial analysis" *per se*, this decision seems to go against even their ideas concerning the nature of "tool kits" and "functional variability" (e.g., Binford 1978b). Moreover, the assumptions underlying Factor Analysis are inappropriate until certain characteristics of the data have been demonstrated (Cowgill 1968). In particular, the use of Pearson's *r* coefficient of correlation assumes that variable distributions are normal (clearly, an unfounded assumption when percentages are used), and the calculation of eigenvectors within the correlation matrix assumes that a linear model is relevant (Doran and Hodson 1972; Rummel 1970; Speth and Johnson 1976). Such technical problems compound the more basic presumptive shortcomings of the Binford's study. Unfortunately, many methods applied to the search for artifact class associations share these kinds of unwarranted and unfulfilled assumptions.

Freeman (1978) also uses Factor Analysis in a search for artifact class associations at the Mousterian site Cueva Morin in northern Spain. In contrast to the Binfords, Freeman incorporates locational information in his analysis. Using informally defined "areas" within the site surface that, he argues, have behavioural relevance, he analyzes artifact class counts for the areas. Even though Freeman has the same problem with the assumptions underlying Factor Analysis as the Binfords, his association study retains spatial information because the units analyzed have locational relevance. Still, his spatial units are informally defined and have the usual problems with evaluation and replicability.

Quadrat Techniques. Many methods for examining spatial associations are based on quadrat units. Most of these techniques involve comparisons between observed artifact frequencies and modelled random distributions. Some compare the observed counts to other artifact classes to determine if coincidence in the distributions is present over space. For example, Dacey uses a chi-square test on a contigency table recording the number of cells in a grid containing members of two classes, one of either class, or neither class (1973). In this technique, significant chi-square values indicate either separation or association between the classes but cannot distinguish between these two types of relation.

As has been discussed, a basic problem with most quadrat techniques, both for pattern recognition and for association, involves the imposition of an arbitrary grid over the distribution under study. It has been noted (e.g., Doran and Hodson 1972; Hodder and Orton 1976) that grid position relative to extant patterning severly effects the outcome of quadrat-based analysis. Because patterns are not known *a priori*, a grid may easily be positioned so as to bisect concentrations of material. This has the unfortunate effect of "creating" dispersion where concentration is really present by filling several cells with frequency values. In extreme cases, this results in Type II error, and a null hypothesis of random patterning may be accepted in statistical tests when the distribution is clustered. There is, at present, no acceptable method for determining if association is present between random patterns, so that quadrat methods run the risk not only of mistaking clustered for random distributions but also of failing to detect associations among artifact classes when they exist. All of these problems can arise due simply to the initial positioning of the collection grid.

Quadrat methods tend to address the problem of spatial clustering and association on a global scale--over the distribution as a whole. This has two results. First, artifact class associations must occur over the entire site surface to be detected. Local associations among classes and the possibility of polythetic activity assemblages are missed. Second, because a single statistic is produced to characterize the degree of inter-class association, the location of such association is lost. This is compounded in methods that use raw artifact counts as the basic data because certain areas of a site surface (refuse dumps are an example) may contain so much of the material present at the site that they "dominate" the association computations. When the goal of analysis is to identify activity-specific tool kits, this becomes an important problem. Artifact classes may have a global association indicated because of the overwhelming influence of refuse dumps on statistical calculations. Such an association may have nothing to do with activity areas or even global patterns but will be interpreted that way since the research design usually defines association as having this kind of meaning.

Not all quadrat methods have these kinds of failings. One technique, developed specifically to address archaeological problems such as these, was proposed by Hietala and Stevens (1977; see above). This method involves the pairwise comparison of quadrat maps using probability models for examining artifact class association over space. The technique uses a series of nonparametric statistics for comparing among maps (chi-square tests, Kendall's <u>tau</u>, loglinear models depending on the kinds of association being considered). The method is promising and produces informative results, but it suffers in two ways. Because it is quadrat-based, the problems with initial grid position discussed above apply here. A second problem involves scaling of data to the least informative class and was also discussed earlier. My own experience with the technique would suggest that this is an important problem, since the degree to which cells are combined into larger units to satisfy statistical requirments severely affects the outcome of analysis. For example, if two artifact classes (one abundant, one rare) are compared based on a meter-square grid, the cells for <u>both</u> classes must be combined in the same way until expected cell frequencies for the rare class are sufficient for statistical testing. What constitutes a sufficient cell size is a matter of some debate. Blalock (1972) says that a minimum value of one is sufficient. Siegal (1956) wants an expected value of five for each cell. I performed Hietala and Stevens' analysis on data from Le Flageolet using both minimum expected values. In the test where one was considered sufficient, there was no association among the artifact classes. In the test where five was considered sufficient, the classes were associated. The lumping of cell units in Hietala and Steven's technique almost always produces fairly large spatial units for comparison (that might obscure smaller-scale pattern association), and analytic results may be conditioned by statistical decisions more than is desireable.

Point Pattern Techniques. Examining spatial association among artifact classes using point pattern data is not well developed in archaeology. Application of this kind of approach includes spatial autocorrelation (Hodder and Orton 1976: 175-83), various forms of nearest neighbor analysis (*ibid*: 202-7), and some forms of local density analysis (Johnson 1976). Spectral analysis (Graham 1980) has been used in archaeology to detect spatial patterning, but its meaning is unclear given the assumptions underlying this astronomical technique. Random walk simulations have also been applied to point pattern data to identify spatial associations among artifact classes (Hodder and Okell 1978). Most often, point pattern analyses concentrate on identifying spatial units that are used to search for pattern association with other statistics. The use of point pattern techniques exclusively in the pattern recognition phase of spatial analysis seems reasonable, since there is no reason to assume that a single technique should or could be used to carry out all aspects of what is a complex analytic problem. The approach favored by many spatial analysts, and that taken here, involves the use of several techniques designed to address the data and problems relevant to each stage of inquiry. As will be seen, this approach comprises the use of *k*-means for pattern recognition and the analysis of association among patterns using different techniques but based on *k*-means results.

It is clear that a search for pattern association is only as useful as are the units being compared. I have already argued that the *k*-means procedure provides useful units. This will be illustrated shortly. The approach to pattern association employed here uses these units within an heuristic analytic framework. Because of this, both informal and formal (i.e., statistical) methods will be used in attempts to detect pattern associations.

PATTERN ASSOCIATION: THE HEURISTIC APPROACH

All of the previously applied techniques for identifying spatial associations among artifact classes have been guided by a concern with defining "tool kits". This is not the goal here. Thus, the concept of spatial association upon which method selection is based is different. The goal of pattern recognition here is to characterize material distributions in space so that they can be assessed in terms of homogeneity. As was discussed in Chapter 2, homogeneity is a relational variable linking variety in content of material accumulations with relative abundance at specific locations. As a first step, it seems reasonable to express the composition of accumulations in terms of subsets--classifications--of the material they contain. In this study, cluster content will be expressed by reference to a series of artifact classes defined by production technology. Abundance reflects the amount of materials present in locations and will be expressed here by the absolute number of objects in an accumulation. Content will be expressed by the proportional representation of distinct classes at a given locus. Thus, association can reflect classes that go together over space in terms of their cooccurrence in clusters and in terms of similar abundance profiles. Homogeneity will be assessed in terms of <u>how many</u> associations are present over a site surface <u>and how</u> consistent associations are with changing material abundance.

In this conception, the assessment of pattern association has two aspects. The first involves the identification of accumulations where many or a few artifact classes are grouped in space. The second comprises separate analyses of content and abundance over space and the synthesis of these into studies of homogeneity. Data produced by *k*-means analysis of individual artifact class distributions solve the first problem. Statistical examinations of these data using a series of nonparametric techniques allow the second problem to be addressed.

Heuristic Zone Definition

In attempting to define associations among artifact classes (based on variety in content) the run statistics produced by *k*-means can be used to generate maps of optimal clustering solutions. Solutions for a variety of classes can even be mapped together at similar spatial scales. Visual inspection of the overlay maps that result from mapping more than one class solution can be easily carried out to determine if there are general areas within a site--accumulations--where centroids are concentrated and where RMS radii overlap. Figure 21 in Chapter 4 represents such an overlay map generated from ten separate *k*-means runs. The decision of how much centroid concentration and RMS overlap constitute an "accumulation" is an intuitive one. The *k*-means maps are used in heuristic fashion, so that basic data description is accomplished mechanically and intuition operates on reduced and clarified patterns. As will be seen in the studies that follow, accumulation definition based on overlay maps is usually quite straightforward.

A brief digression concerning the use of informal techniques at this stage of analysis is appropriate. During the course of research, attempts were made to quantify spatial coincidence among *k*-means maps. These attempts continue. One approach involved comparing maps by determining the area of overlapping circles. At first, this seemed a fruitful approach; computations were not complicated, requiring only measures of distance among cluster centroids and the use of RMS to calculate cluster overlap. A basic problem arose with this method, however. As was discussed earlier, the relation between the RMS radius and the density and dispersal of materials associated with a cluster is complex. A number of factors affect the RMS in a variety of ways, and the nature of these influences must be studied in detail before a useful association statistic can be proposed. Attempts at quantifying inter-class associations based on *k*-means analysis were (temorarily, at least) abandoned in favor of more direct and simple techniques.

Thus, a decision was made to use *k*-means analysis heuristically and to rely on intuitive methods for defining areas of material accumulation over a site surface. These would then serve as units for examining artifact class associations over space and homogeneity among accumulations. In each of the sites analyzed as case studies, a series of *k*-means runs are performed on individual artifact class distributions. These are combined on overlay maps at different optimal clustering levels based on centroid and RMS statistics. Maps are inspected in order to define spatial units that will be called "zones" and that correspond to discrete areas of material accumulation. Specifically, *zones* will be defined based on overlay maps when *k*-means cluster centroids are concentrated and RMS circles overlap. Important characteristics of zones are that they are heuristically defined and have spatial relevance.

In practice, zones will be defined in several ways. One condition that will result in zone definition is many cluster centroids concentrated in a small space with extensive overlap among their RMS circles. A second condition that will result in zone definition is the clear spatial separation of a set of centroids from surrounding zones, i.e., little RMS overlap with member clusters of other zones. Zones are also distinguished when there is substantial overlap in RMS within two groups of clusters and less overlap between them, even if the centroids are relatively close together. Finally, a zone will be defined when a set of clusters is bounded by a series of other zones having the above characteristics, even if the set itself is diffuse. Zones are always comprised of more than one cluster. Clusters that are separate from all others in space are termed isolates.

It should be evident that this approach to zone definition will produce spatial units that do not necessarily represent the same kind of relation among artifact class clusters. However, in their variability, these units have the advantage of retaining any local variation in spatial patterning. The method's primary disadvantage is the subjective nature of zone definition. As the method is applied, the decision-making process will be discussed in light of this subjectivity, and the criteria for specific zone boundary definitions will be made explicit so that the nature of the heuristic analysis will be clear.

Global Association among Artifact Classes

Using zones defined in this way, a series of techniques will be used to determine if artifact classes associate with others over space and to examine homogeneity within and between accumulations. The first method employed after pattern recognition is designed to identify associations among classes. Objects within a zone are counted by class, the counts are converted to ranks within classes, and correlation analysis is performed on the resultant rank matrix. Because, as will be seen, k-means analysis identifies several optimal solutions for every distribution analyzed, correlation is assessed at multiple scales of spatial refinement.

Rank order correlation is used here for several reasons. The most important involves potential problems arising from the use of correlation measures on raw counts (cf. Speth and Johnson 1976). The best known correlation measures are parametric and assume that the variable distributions under study are normal. For archaeological data, this assumption may not hold and must be demonstrated. Nonparametric statistics, including rank-order correlation, do not make this assumption of data distributions (Siegal 1956). Most nonparametric statistics involve conversion of data to ordinal scales, and ranks represent this level of data expression. Nonparametric statistics have the added advantage of relative robusticity even when samples are small (Blalock 1972). As will be seen in the next chapter, this is the case for artifact counts in many of the spatial zones defined during the archaeological case studies.

To determine if individual artifact classes associate over space, Spearman's *rho* statistic for rank order correlation is calculated between each pair of artifact classes. The basic strategy is as follows:

1) *K*-means analysis is performed on separate artifact class point distributions.

2) All optimal clustering solutions between one and fifteen clusters are identified by inflections in log(%SSE) plots for each artifact class.

3) Overlay maps for all artifact classes are produced at each identified clustering scale. The solutions for all classes comprising the fewest clusters are mapped at a *low scale solution*. Solutions with the greatest number of clusters are combined for a *high scale solution*. Middle-range solutions are mapped together at the *medium scale*.

4) Zones are defined for each overlay map.

5) The number of artifacts in each class are counted for each zone.

6) Raw counts are converted to within-class ranks; for example, the zone with the most endscrapers has a rank of one, the zone with the second largest endscraper population has a rank of two, and so forth.

7) Spearman's *rho* is calculated for a rank matrix where artifact classes are rows and zones are columns. Correlations are computed between pairs of artifact classes.

8) *Rho* values are assessed for significance.

Spearman's rank order statistic is simple to calculate:

$$\textit{rho} = 1 - \frac{6\left[\sum_{i=1}^{N} D_i^2\right]}{N(N^2-1)}$$

where N represents the total number of cases (zones) and D is the difference between ranks for a given case. The value thus obtained can range from +1 when there is perfect aggreement among all rank values for two classes (here, perfect association), through 0 when there is no relation, to -1 when the rank values are perfectly opposite (the classes are separated in space). Significance can be assessed for each correlation statistic based on the sample size involved and the magnitude of the correlation value. Significance is here examined using a normal deviate and is computed as follows:

$$\underline{Z} = \frac{r-0}{1/N-1}$$

$\underline{Z}$ is evaluated using a $\underline{z}$-table at predetermined significance levels. The significance level employed here is ($p < .05$), although both higher ($p < .01$) and lower ($p < .1$) levels are sometimes discussed. For all correlation tests, the null hypothesis being tested is:

> H0: The observed correlation coefficient is not higher that would be expected due to chance given the sample sizes involved.

Using this strategy, there is a 5% change of Type I Error, rejecting the null hypothesis when it is true. Conversely, we can be 95% sure that correlations defined as significant represent systematic coincidence in rank values between artifact classes.

The use of rank order correlation allows associations among artifact classes to be identified on a global level, i.e., over an entire archaeological surface characterized as a series of accumulations. Such a global measure is useful, since it gives an indication of site-wide relationships. Moreover, because several different refinement scales are analyzed for each site, changes in global relations among artifacts can be examined through the sequence. However, local association among classes is not assessed using this correlation technique.

Local Association among Artifact Classes

The problem of local association in space is somewhat more complex than global assessment of correlation. To determine of there are associations among classes that might be present in a few locations across a site surface, a technique is required that retains at least some of the individuality of specific zones. Such a technique must also allow zones of differing richness to be compared without rich zones overpowering sparse ones because of their numerical superiority. Unfortunately, no statistical methods exist that fulfill these requirements. As a preliminary attempt at local association assessment, an heuristic and intuitive approach is taken here. The examination of local variation in artifact association will be referred to as "content analysis" in the studies to follow. This specific term is used because it is the variation in zone composition at various levels of resolution that is of concern. No formal method is employed in comparing zone contents. Instead, counts for individual artifact classes are standardized to eliminate "noise" produced by relative abundance within and between zones, and the standardized values are displayed and evaluated visually.

The standardization procedure utilized here is not new to archaeology, but it has not been used often. Binford applies the technique to compare between bone assemblages showing variability in the number and kind of animals present (1977, 1978a, 1981). He uses the method to "iron out" abundance related counting problems frequently encountered in comparing faunal data. It is for similar reasons that the same standardizations are employed here.

Standardization of data is accomplished using a rather simple formula. In two different dimensions--by row (artifact class) and by column (zones)-- matrices of raw artifact counts are expressed as percentages of the maximum value in a dimension. Two standardized matrices are thus produced from each raw count table: a <u>class maximum</u> and a <u>zone maximum</u> matrix. The first four steps <u>of the standardization</u> procedure are identical to those employed in correlation analysis (see above). Succeeding steps are as follows:

5) To produce class maximum values, each cell in the matrix is divided by the maximum value in its row and the result is multiplied by 100. A new matrix is obtained representing the percentage of the richest occurrence of an artifact class contained in each zone. Class maximum matrices have the following characteristics: (a) there is a maximum value of 100 present in every row, representing the location of maximum abundance for the artifact class, and (b) other row values represent the percentage of the maximum value present in each zone. Class maximum standardization reduces the effects of relative abundance within artifact classes so that the distribution of abundance itself can be examined over space. Variation in class maximum values increases as classes are concentrated in a few locations, so that high values in one or two columns (zones) insures low values in others.

6) Zone maximum standardization is performed within columns of the raw data matrix rather than within rows.

The same calculations are carried out. All cell values in a column are divided by the largest cell value in the column, and the result is multiplied by 100. Each column has at least one maximum value of 100. This matrix expresses zone contents as percentages of the richest class in the zone. Using zone maximum standardization, zone contents can be compared with variation in overall zone richness controlled. In other words, rich zones can be compared directly to sparse ones.

7) Results of standardization are displayed using techniques described below.

Class and zone maximum standardizations are useful because they provide two different perspectives on the data and retain local differences among zones. Sample size variation is controlled in two dimensions, yet the effects of variability in material abundance can be examined over the site surface. In addition, local variation can be examined through comparisons among spatial units on a single measurement scale.

It should be pointed out here that, while standardized data provide some measure of comparabiltiy among zones of different richness, the effects of stochastic variation on very sparse zones (those with few artifacts) may be great. This is especially the case for zone maximum values in sparse zones; the addition or subtraction of a few artifacts of a given class might dramatically alter the zone profile if there are few artifacts altogether. Class maximum values can also be affected by stochastic processes of this kind for artifact classes that are poorly represented in the site as a whole. However, this is less of a problem in the studies that follow, since most artifact classes have reasonable sample sizes. The potential effects of stochastic variation should be kept in mind as the reader progresses through the case studies. In particular, visual screening of content profile graphs must be carried out with these potential effects in mind. This cannot be stressed too strongly and will be considered again later as specific analysis are discussed.

To make consideration of content profiles easier, a display method will be employed that allows standardized values to be represented graphically. The illustrative technique was originally borrowed from geologists, who use it to depict the composition of sediment samples. Archaeological usage has been in illustrating assemblage composition (e.g., Delporte and Mazière 1977; Rigaud 1978, 1982). The display involves a circular graph. A circle, the circumference representing a maximum value (here 100% of the possible maximum), is segmented by equidistant radii. The radii are assigned to artifact classes and ruled from zero at the center to 100 at the perimeter.

The standardized value for each artifact class is located along the corresponding radius, and these points are connected. A polygon within the circle is produced in this way. Peaks in the polygon represent classes that have higher standardized values than others. To facilitate comparisons, specific artifact classes are always located on the same radius on all graphs. Thus, graphs are comparable between levels of spatial refinement as well as between zones at the same refinement scale. At every clustering solution level, a zone defined on the overlay map will have two content profiles and two circular graphs. One profile depicts class maximum standardized values for the zone; the other illustrates zone maximum standardized values. For the Aurignacian levels from Le Flageolet I, the artifact classes employed in the analyses are the same. Therefore, content profiles also allow comparison among archaeological levels at that site.

Examining Homogeneity over a Site Surface

Finally, to assess homogeneity in zone contents at a particular spatial scale and to examine changes in homogeneity between scales (in short, to provide the means for evaluating the models presented in Chapter 2), the nonparametric correlation matrix produced at each scale of spatial resolution will again be used. At a given solution level, homogeneity will be characterized by the percentage of all possible correlations that are positive and significant. This percentage can be plotted over all solution scales for a distribution from largest to smallest, in order to see changes in the pattern of zone content and abundance. By expressing homogeneity with this percentage, the amount of potential artifact class relationships that actually exist can be examined and compared between levels. For example, a percentage value of 85% indicates a very homogeneous series of zones since many artifact classes are present in similar relative abundance over the entire site surface, i.e., there is not much variety in zone content that cannot be attributed to sample size. A low percentage value is expected when only certain artifact classes go together in specific places while other classes are independent or have other pattern associations. This condition implies a variety of zones in terms of their contents. Both abundance and content are accounted for in this percentage figure by using rank order correlations (reflecting abundance) by artifact class (reflecting content).

OUTLINE REVIEW OF THE ANALYTIC SEQUENCE

In summary, the methods of analysis discussed in this chapter will be applied in identical sequence to the spatial distributions from Pincevent and two archaeological levels from Le Flageolet I. A brief review of the techniques to be employed and the proposed analytic sequence in outline form would seem appropriate at this point:

A. Stage I Analysis: Pattern Recognition.

1. *K*-means cluster analysis of individual artifact class distributions.

2. Examination of the log(%SSE) plots for each clustering run to identify optimal solutions for each artifact class distribution.

3. Examination of maps produced with *k*-means run statistics (centroids and RMS) for individual artifact classes to assess patterns within and between class distributions.

B. Stage II Analysis: Pattern Association, Zone Definition

1. Production of overlay maps at a variety of solution scales using *k*-means run statistics for individual class distributions. Heuristic definition of zones for each map.

C. Stage III Analysis: Pattern Association, Content Analysis

1. Production of artifact class frequency counts by zone for each *k*-means solution level. Artifact classes form the rows and zones form the columns in the resulting frequency matrix.

2. Analysis of global associations among artifact classes. Raw frequencies produced in A:(1) replaced by row ranks. Spearman's *rho* used to define correlations among artifact class pairs. Correlations tested for significance. Correlation analysis is carried out for all solution scales defined in B:(1).

3. Heuristic examination of local variation in zone content. Raw frequencies are standardized in two dimensions: as percentages of row (class) maximum cell values and of column (zone) maximum cell values. Circular graphs produced to display varieties of content profiles. Raw frequency tables produced in B:(2) are treated in this manner for all solution scales.

4. Examination of zone homogeneity. The percent of each correlation table produced in B:(3) comprising significant positive relations among classes is calculated for each solution level. In all computations, ($p < .05$) is considered significant.

5. Examination of changing homogeneity with changing spatial refinement. Percentage values obtained in B:(4) are plotted for all solution scales defined for a single archaeological distribution. These plots are compared to models discussed in Chapter 2.

C. Interpretation of Results.

1. All analyses performed for an archaeological level are discussed and inferences drawn from results.

In the following chapters, this analytic process will be applied to artifact distributions from three Upper Paleolithic occupation levels. In all of these case studies, relationships between the results of spatial analyses, formal and informal, and other characteristics of the archaeological record under examination will be discussed. In order to make the results of spatial analyses more understandable, and in order to provide heuristic means for evaluating results at each analytic stage, elements of archaeological context will be used in making inferences about quantitative observations. Elements like hearth features, topography at the sites under investigation, and the overall configuration of available space will be considered in assessing and interpreting spatial analyses. For this reason, general discussions of the two sites, including previous analyses and interpretions, will preceed each detailed spatial analysis. These discussions will be referred to frequently during the presentation of spatial investigations and are integral to the heuristic approach.

TERMINOLOGY

In concluding this chapter on analysis, a brief consideration of the terminology employed in case studies is in order. Terminology is an important element in the investigations which follow. Words will be used rather precisely, and the meanings implied here may differ slightly from normal usage. Any such differences involve more exact meanings than may be current. Many of the terms have already been discussed in this and the preceeding chapters. However, other terms require definition as well, and these often refer to aspects of the archaeological record that will be integrated with quantitative results.

Terms Referring to the Archaeological Record

The first set of terms to be defined refer to general characteristics of the archaeological record and the data to be analyzed. The basic spatial unit for analysis, comprising all artifacts occurring within a distribution, is the surface. This term corresponds to the French sol d'habitat and is assumed to represent an area inhabited or used by humans during a single "occupation" of a site. A synonym for surface is floor. Occupation here refers to an analytic unit: the finest, archaeologically discerned stratigraphic entity. This term is synonomous with archaeological level and may, in fact, represent several brief periods of actual site use which cannot be distinguished by archaeological means (e.g., sedimentology, lithic conjoining, microstatigraphy, etc.).

A surface may contain features or the remains of human activity that involve multiple objects (in essence, were constructed of parts) and that are restricted in spatial extent yet cover more of the surface than does a single artifact. A specific type of feature related to the use of fire (virtually all of those considered here) is a hearth. Hearths are frequently constructed of igneous or metamorphic rocks imported to sites for this particular use. The part of a surface immediately surrounding a hearth will be referred to as an area; area is a general term and does not refer to analytically defined spaces. Another term used to refer to generalized parts of a surface is sector. Sector will be used to designate an extensive part of a site surface separated from other parts by contextual, e.g., topographic, characteristics.

Terms Referring to Aspects of Spatial Analyses

Terms that are used to describe aspects of spatial analysis have been discussed, but they are reiterated here for reference. Analytic terms refer both to characteristics of *k*-means analysis and to more general concepts derived from specific studies.

The most basic term used here is pattern. Pattern will refer to an organized arrangement of materials over a surface. Patterns can be described, for example in terms of clusters, but while clusters are defined analytically, patterns are present in the archaeological record. They are empirical phenomena. A cluster refers to a single spatial unit defined by *k*-means analysis of an artifact class distribution in a single solution. A cluster has a centroid expressed as coordinates in space and defined by averaging the _x_ and _y_ coordinate values of cluster members. A cluster has a _size_ defined by the _RMS_ radius (a radius from the centroid within which 66% of all cluster members are located). A _large_ cluster has a long RMS, but this usually means that it is a dispersed concentration of points. A _small_ cluster has a short RMS radius and usually represents a compact and dense concentration of points. The number of artifacts making up a cluster is its _point frequency_. Thus, a point frequency refers only to a single artifact class.

Several terms refer to the _k_-means clustering process. As was discussed earlier, every _k_-means run produces a series of _SSE_ statistics by summing the squared distances between all points in all clusters and their nearest centroid at each cluster solution level. This statistic is used in searching for optimal solutions for a given distribution. The SSE is transformed to produce the _log(%SSE)_ which is a scalar measure of the change in clustering efficacy from solution to solution. Inflections in a plot of the log(%SSE) against solution levels indicates optimal descriptions of the data distributions. As the _k_-means procedure finds cluster solutions from one to as many clusters as the user stipulates, the process passes through a series of _cluster levels_. At each level (i.e., one, two, to _k_), the procedure finds a _solution_, representing the "best" division of the distribution into that many clusters, by minimizing the SSE. _Optimal solutions_ are those that the log(%SSE) plot identifies as patterns in the distribution. In the discussions that follow, the adjectives _low_ and _high_ are used to qualify "level" and "solution". These terms relate to the _k_-means sequence: lower levels are those formed early in the run and are comprised of fewer clusters, and higher levels are formed later and are made up of more clusters.

For each _k_-means analysis of an artifact distribution, several optimal solutions are expected. These can be organized, for purposes of comparison among distributions, along a _scale_ from low level to high level. One artifact class distribution may have optimal solutions when, for example, two and five and eight clusters are distinguished. A second artifact class may have optimal solutions at three, five, and eleven cluster configurations. In comparing these two classes, low scale solutions include the two cluster pattern from the first and the three cluster solution from the second. While the absolute number of clusters is slightly different, the scale relationship of the cluster levels being compared for the two distributions is the same. In this example, a medium scale can be distinguished with a five cluster solution from the first class and a five cluster solution from the second. High scale solutions also occur in both class distributions.

To compare between individual *k*-means runs, clusters at similar scales are mapped together using centroid and RMS statistics, producing an *overlay map* corresponding to a single scale of spatial refinement. Based on these maps, *zones* can be defined heuristically when centroids are concentrated and/or RMS radii overlap among clusters or when more diffuse clusters are bounded by compact zones. Each surface may contain few or many zones at every solution scale. Individual pieces or clusters that are so far away from all other clusters as to have no RMS overlap are termed *isolates*. It should be noted that individual artifacts may be parts of zones. Also, a zone is an analytic concept; it may thus occur in an area, i.e., near a hearth, but cannot alone compose an area.

Finally, once cluster zones and isolates are identified at a variety of scales, they must be characterized for further analysis. A zone can be described in terms of its *content*, referring to the point frequencies for all artifact class clusters composing the zone. Content for a given class is calculated by summing point frequencies for all clusters of a class in a zone, and content includes zero values for classes having no clusters in the zone.

With these definitions in mind, we turn now to applications of heuristic spatial analysis. The first case study examines the artifact distributions at the Upper Paleolithic site at Pincevent. Then, two occupations uncovered at the Upper Paleolithic rockshelter site Le Flageolet I are analyzed using the same procedures. These studies are compared at the end of this volume in terms of depositional characteristics inferred from spatial analyses and evaluation of the models presented in Chapter 2.

CHAPTER 4

PINCEVENT SECTION 36: LEVEL IV(2)

INTRODUCTION

The Magdalenian occupation site at Pincevent (Seine-et-Marne) is one of the most interesting Upper Paleolithic localities known in Europe. Extensive analyses have been performed and published by the excavators (Leroi-Gourhan and Brézillon 1966, 1972). Because much of the analysis that follows depends heavily on the interpretations presented by Leroi-Gourhan and Brézillon, it seems appropriate to introduce the site, as described in their monograph, in some detail. This discussion will serve to acquaint the reader with the site itself and to outline data and interpretations that will be relevant to the present study.

Pincevent is that rarest of archaeological finds, a series of well-preserved Paleolithic occupation surfaces. The area designated "Section 36" has yielded five archaeological levels within a deep sequence of flood silts and varves. Rich Upper Paleolithic surfaces occur in Level IV, which seems to extend over one and one-half hectares. Preservation at Pincevent resulted from rapid burial of cultural deposits by gentle flooding. It seems that the Magdalenian occupations of the site occurred during a period of considerable river movement. The site was occupied when dry and quickly buried as waters rose. Lenses of gravel and sand are found within the silt sequence, indicating that the Seine moved over the location between occupations. The two meters of flood silts, frequently discernible as varve sequences, suggests that low transport strength characterized the deposition. Deposits were rapidly protected from the ravages of weather and scavengers, yet their spatial integrety seems to have been little disturbed during inundation.

Two microstratigraphic occupation levels can be discerned within Level IV. Level IV(4) is an early and not very dense layer that may relate to the Habitation 1 locus (Leroi-Gourhan and Brézillon 1966). A much denser layer, IV(2), occurs higher in the silts, and this was excavated over 600 square meters of surface. Twenty-three features, over 94,000 kg. of flint, the remains of at least 43 reindeer (*Rangifer tarandus*), and masses of fire-cracked rock, manganese and iron oxides, etc., were recovered from the IV(2) occupation surface. All finds are mapped and registered, providing a complete and rich data base for detailed spatial analysis.

The monograph on Pincevent Section 36 begins with Brezillon's discussion of the lithic materials. He notes a rather high percentage of backed bladelets in the assemblage, and a low burin count. Percoirs (awls) also occur in greater numbers than expected for this Magdalenian phase.

"Topographic Analyses", or the examination of spatial distributions, is then undertaken for all materials recovered. The spatial arrangement of each material class (e.g., fire-cracked rock, ocre, burins, antlers) is examined separately and the findings integrated into an interpretation of the site.

During these analyses, a series of observations are made by Leroi-Gourhan and Brézillon that will be relevant to this study. First, the twenty-three features uncovered seem to be grouped into four types. The three largest features are concentrations of ash, burnt bone, and fragmented rock within shallow, slab-lined depressions. These are interpreted as the remains of central hearth structures (fr. foyers) and seem to be the focus of much surrounding deposition. Flat, unlined hearths occur that are filled with gravel and small fragments of rock. "Gravel hearths" are concentrations of carbon fragments covered or blended with gravel. Finally, small depressions containing little material are designated as "small cup hearths" and are frequently associated with large flat limestone slabs positioned within one to two meters.

Another series of observations concerns the distribution of red ocre staining relative to central hearth positions. Heaviest staining occurs to one side of all three central hearth features and coincides with the area of densest stone tool concentration. Ocre staining is discontinuous between central hearths, producing three general areas of staining with clear areas between.

Lithic waste also shows characteristic spatial concentration. Debitage tends to be concentrated in small piles (ibid: 101), and these piles occur in the areas of the central hearths and near a small hearth in unit V101 (see Figure 6 for this and following locations). According to Leroi-Gourhan and Brézillon, the general impression of the debitage is that no "workshop" concentrations of waste are present. They suggest that human alteration, selection, and movement of lithic materials occurred at all locations on the site. In other words, specific pieces were moved to particular locations for use, but further selection, manipulation, and transport occurred afterwards. It is possible that entire knapping and use "events" happened, and the results were removed from one location in containers or ground-cloths of skin to be redeposited elsewhere. Debitage tends to be concentrated, with fire-cracked rock, on one side of the three central hearth features. The distributions of ocre (and stone tools) coincide on the opposite side. This pattern is repeated at all three central hearth locations.

Further evidence for redeposition of lithic materials after production involves flint cores. Cores occur at the far margins of refuse masses associated with central hearths and are rarely found in association with debitage piles. At the small hearth located in unit I101, which shows no characteristics of a central hearth, cores and debitage are found together. Even here, however, the two artifact types are separated by the hearth itself. A general lack of other material types at I101, i.e., bone, rock, ocre, etc., argues for its particular nature. The excavators assert that, "its composition seems to indicate that it was only used once" (op. cit.: 116), which suggests a functionally specific purpose.

To Leroi-Gourhan and Brézillon, observations concerning the general distributions of serveral artifact categories--fire-cracked stone, cores, flint debitage, ocre--reflect the repetition of a single pattern three times. Three large, slab-lined hearths act as the foci for all materials. Tools made of stone and red ocre are concentrated on one site of the hearth. Bone debris and fragmented rock form refuse masses on the other side of the hearth from tools and ocre. Debitage seems to occur in small pockets within the debris piles, while flint cores occur at the outer margins of the bone/rock/lithic masses.

Interpreting these general observations, the excavators suggest the presence of three habitations centered on the three focal hearths. They refer to these, after the excavation unit where the features occur, as V105 in the northwest, T112 in the northeast, and L115 in the southeast. The authors conclude that the hearths were the foci of "habitation" units since they repeat essentially "identical spatial patterns of fire-cracked rock, ocre, and piles of lithic material (knapping waste and cores)" (Leroi-Gourhan and Brézillon 1972: 116).

Leroi-Gourhan and Brézillon analyze the spatial distributions of individual artifact classes in terms of the tripartate divisions identified above. Stone tool classes have patterned relationships to the central hearths. Burins, which have a low count for the site assemblage as a whole, seem to be widely distributed over the entire excavated surface. In habitations V105 and T112, burins tend to occur in "domestic areas" in association with ocre and other tools, and they "thus seem to have been abandoned in place" within these areas (p. 124). Burins also occur in refuse masses associated with each central hearth but in much lower density. Burin distribution in the area of L115 is "non significant" (p. 125), i.e., they appear to be randomly scattered throughout the southeast portion of the site. Burins also seem to associate with flat hearths G115, C114, and P102.

Backed bladelets have a different distribution. These tools seem to be concentrated near or within the three central hearth features. Where they occur, they are abundant in a relatively small area. Five areas are the foci for backed bladelets: within the hearths, in the refuse mass before V105, in a small zone northwest of V105 (designated X-Y,102-103), in an area between the two northern hearths (Q-R-S,107-108), and west of flat hearth G115.

Percoirs and micropercoirs have a high frequency in the assemblage as a whole. The most striking characteristic of their distribution is concentration around hearth T112. Elsewhere, they tend to be concentrated at central hearths in association with ocre and other tools.

Endscrapers have a somewhat different general distribution than other tool classes. They tend to be scattered over the site, occurring in domestic and refuse areas close to the central hearths and near certain outlying features. There are no dense concentrations of scrapers, although six of these tools are dispersed within two meters of feature Q111 which has no other stone tool associations (p. 137).

The distribution of bone materials is generally similar to that for stone. Dense masses occur near V105 and T112, where they occupy the "refuse" side of the hearths along with fire-cracked rock and debitage. Around L115, bone tends to be less abundant and more dispersed.

In looking at the distributions of specific bones, Leroi-Gourhan and Brézillon note that antler fragments tend to be located only in refuse dumps near focal hearths, and larger antler pieces are found at the external edges of refuse piles, i.e., furthest from the hearth itself. All large antlers appear to have been worked.

Dental materials provide an estimate of minimum number of individual animals present in the hearth refuse piles. The two northern habitations contain the remains of around twenty animals each, while L115 contains a minimum of only two animals. Spatially, teeth and mandibular/maxillary fragments tend to concentrate in refuse piles associated with the central hearth features. Few dental remains are associated with L115, and these occur at a 2 m distance from the feature. The authors argue that the concentration of dental materials near the focal hearths indicates that animal heads were processed at the hearths:

> It might be suggested that preparation of the head for consumption occurred in the immediate proximity of the three principal hearths... [1972: 172].

Since caribou hunters rarely transport the head from kill location to base camp sites, the presence of heads and their on-site processing would indicate either a nearby kill/cache location or something of a production crisis in the hunting population (see Binford 1978: 47-90).

Other bones (long bones, metapodials, ribs, vertebrae, tarsals, phalanges, and sesamoids) are all more or less concentrated in refuse areas near focal hearths. Smaller features show limited bone content, although vertebrae seem to be scattered around the excavated area forming little concentrations at some peripheral hearths. Flat bones (e.g., omoplates) are entirely absent from the central hearth zones. They tend to occur at the outer periphery of refuse piles, however, supporting the notion that heads were processed at the central hearths.

As their interpretation of the Pincevent Section 36 deposits, Leroi-Gourhan and Brézillon present a "model" or stylized description of spatial organization within the inferred habitation structures (1972: 254). The central hearth itself is designated as Zone A since it serves as the focus of habitation (Figure 6). Surrounding the hearth is a B zone with two subdivisions. Zone B1 is that area to one side of the hearth where ocre and finished stone tools dominate. A relatively empty area which repeatedly abuts the B1 zones at each hearth is identified as interior space, since the excavators argue that the lack of refuse indicates a covering superstructure over these zones. Zone B1, then, is defined as interior activity space. On the other side of the hearth, Zone B2 represents exterior activity deposits that include masses of bone debris, lithic waste, and fire-cracked rock.

With increasing distance from the central hearth towards the exterior of the inferred structures, a series of zones reflect decreasing refuse deposition. The D zone represents near refuse evacuation space and occurs from one to two meters from the hearth. Here, quantities of bone debris and debitage reflect dumping behavior from the interior side of the hearth. Three to four meters from the hearth, a dispersed evacuation space is designated as Zone E. Zone F, from five to six meters away from the central hearth, represents an area of rare deposition. Finally, Zone G makes up most of the excavated area, and represents areas of rare deposition between the habitation structures.

The evidence and conclusions presented by Leroi-Gourhan and Brézillon as summarized above will serve as a comparative interpretation for the quantitative study

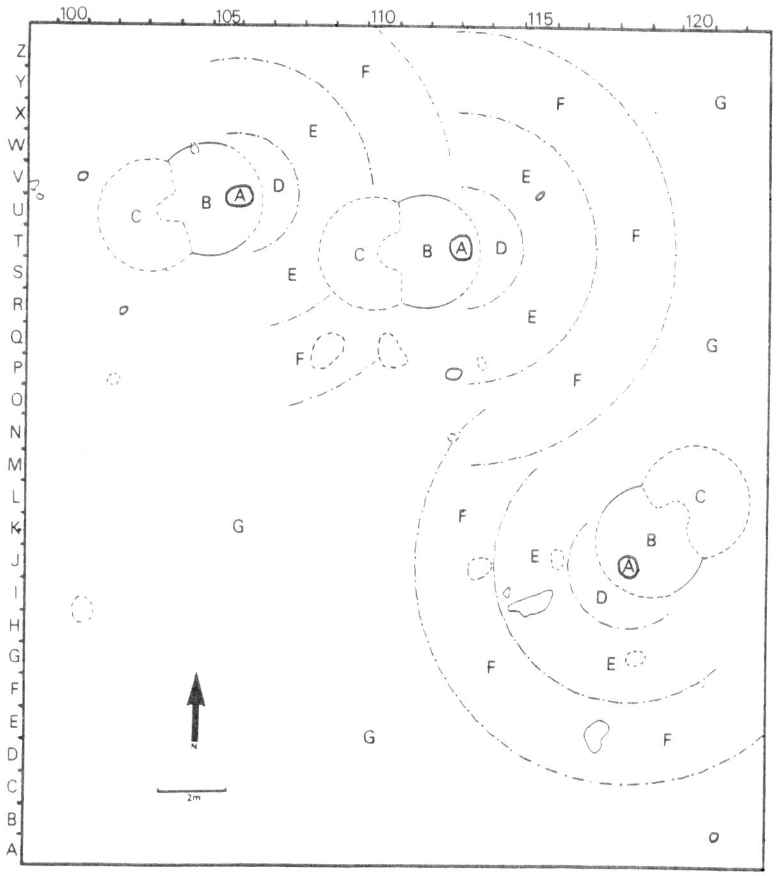

Figure 6. Excavators' Model for Habitation Zones at Pincevent Section 36 Overlaid on Map of Features (after Leroi-Gourhan and Brézillon 1972).

undertaken here. The excavators' conclusions will not be "tested" in a formal sense. Instead, the same data will be analyzed in a quantitative way. Coincidence between intuitive and quantitative analyses would not, methodologically, increase the plausibility of the excavators' conclusions, but confidence in the quantitative method's ability to detect reasonable patterning would be bolstered. Furthermore, quantitatively defined spatial units can be used for finer scale examinations of spatial variation in artifact association, depositional intensity, etc., since they can be characterized in terms of size, shape, and frequencies of artifacts. In other words, they provide explicitly defined analytic units.

Two goals guide the first spatial analysis presented. First, the k-means method will be applied to the Pincevent data, and results will be compared to those obtained by extensive intuitive analysis. Second, if the two sets of results diverge, the differences must be explained. If the two coincide, however, further analyses will be undertaken to examine spatial variability in artifact association.

Description of the Data

Data used in this study consist of point coordinates in two dimensions. Ten data maps for artifact classes are distinguished. These include five stone tool classes: *percoirs* or awls, endscrapers, burins, chert cores, and backed bladelets. Five caribou bone classes include: metacarpals, metatarsals, antlers, mandibles and teeth (considered together), and ribs. In all classes, fragmentary materials are included if they were mapped by the excavators. Spatial coordinates were obtained from the maps published in the monograph.

Classes of bone and stone materials conform to those used by Leroi-Gourhan and Brézillon. For bone, these subdivisions are based on identified body parts. Since only caribou bones were identified and mapped in detail, other animal bones (which make up a very small part of the faunal component at Pincevent) are excluded here. Stone tools are subdivided into general technological classes. Although these divisions are conditioned by their presentation in the monograph, the broad classes provide (1) sufficient sample sizes for individual analyses by class, and (2) groupings that may reflect certain aspects of tool use better than the finer scale traditional types. For example, seven types of endscrapers occur according to the de Sonneville-Bordes and Perrot typology (1956, 1957). Four types of percoirs are counted (Leroi-Gourhan and Brézillon 1972: 52-53), while thirteen burin types are noted. Backed bladelets comprise a single type according to the list, and cores are grouped under a single heading without accounting for morphological variation. The refined types of the classic typological list have been criticized in the past, since the reference of specific types, i.e., to chronological, functional, or stylistic variation, has never been firmly established. The more general classes mapped by Leroi-Gourhan and Brezillon reflect basic techniques of tool manufacture that are consistent in a general way over time and space.

It should be noted that the total number of objects within a class analyzed here may vary somewhat from totals for the site presented by Leroi-Gourhan and Brézillon. This discrepancy results from our data collection technique. Cartesian coordinates were obtained from the published maps using a computer digitizer. All points on the published maps were recorded and checked. Yet an error factor is introduced by this method since data modification or rerecording always reduces accuracy. Objects not present on the published maps, e.g., those found during screening operations at the excavation, are not included in the quantitative analyses which follow. A bias is probably present here against small tool classes, backed bladelets for example, but the recovery techniques employed during excavation ensure that most if not all objects in a class are present on the published maps. Table 1 presents the artifact classes used for the analyses which follow, the number of objects within each class recorded from the published maps, and the Figure and page references from the site monograph.

STAGE I ANALYSIS: PATTERN RECOGNITION

The first stage of the Pincevent case study involves the application of k-means cluster analysis to each artifact category. Each analysis is carried to a fifteen cluster solution. Above fifteen clusters, single object clusters were consistently formed from solution to solution. This characteristic probably reflects a final solution level above which no further progress is made in pattern recognition.

Presentation of the details of each individual run would be long and repetitive. Only some of the runs are discussed when they represent general patterns or when class specific patterning is indicated. As an introduction, the k-means run for mandible and teeth locations is outlined to illustrate the method. Other classes are discussed in less detail.

From Leroi-Gourhan and Brézillon's published map, 219 points were digitized for caribou dental materials. These were undifferentiated as to whether they were whole or fragmentary specimens, or if they were mandibular or maxillary parts.

TABLE 1.

Artifact classes used for Pincevent 36 analyses and references to class distribution maps in Leroi-Gourhan and Brézillon (1972). This Table is reproduced from Simek and Larick (1983).

ARTIFACT CLASS	N	FIGURE	PAGES
Cores	149	66	112-115
Burins	177	70	126-128
Backed Bladelets	691	73	134-136
Endscrapers	132	74	138-140
Percoirs	136	74	138-140
Antlers	164	79	156-158
Mandible Fragments	219	92	168-170
Metacarpals	235	95	178-180
Metatarsals	186	95	178-180
Ribs	366	96	182-184

The first run statistic of interest is the log(%SSE) plot. Figure 7 illustrates that plot, and it shows three cluster levels with good clustering solutions. Using the cluster centroids and RMS measures, three maps can be produced, one for each indicated level. The first of these levels, three clusters, shows clusters positioned near the three central hearth features defined as "A Zones" by the excavators (Figure 8). Centroids are placed in those areas designated refuse deposition zones, i.e., on the proposed "exterior" sides of the hearths. This relationship between quantitatively defined patterning and structural features, at the three cluster level, coincides with the excavators' interpretations.

The six cluster solution (Figure 9) shows four separate clusters formed in the northwest quarter of the excavated area. The two other clusters more or less conform to the initial three cluster solution. Some individual points are lost from these latter clusters in the breakdown and refinement of the northwest distribution. Tables 2 and 3 show the number of mandible fragments present in each cluster at the three and six cluster solutions, respectively. Also listed in these tables are the centroid coordinates and RMS statistics by which Figures 8 and 9 were produced.

It is interesting to note that the richest cluster produced at the three cluster level--the northwest cluster--is the first to be divided. It is partitioned at the six cluster level, while other areas are left as defined earlier. Perhaps the area of highest artifact concentration reflects several smaller concentrations within a limited area. If this is so, then finer-scaled patterning would be indicated within the three cluster general pattern. This finer clustering, moreover, seems to relate the abundance of objects in a general area to the location of smaller concentrations, i.e., a kind of tiered packing of smaller units into larger ones. Other examples of this relationship will be discussed where they occur in other individual artifact categories.

A nine-cluster solution is also indicated in the log(%SSE) plot. Table 4 presents the cluster statistics for this solution, and Figure 10 shows the map produced from these statistics. Rich clusters are located close to the central hearth features again. Other clusters are positioned further away from the three hearths, and seem to be near features designated as peripheral by the excavators. Areas defined as "C Zones"--the interiors of inferred structures, have no cluster centroids. The three richest clusters are those associated with focal hearths, with over 68% of all dental materials. (This pattern also obtains at the six cluster level, where over 83% of all objects occur within the three clusters closest to the hearths.) The northeastern and southern clusters, which remained intact at the first two clustering levels, are subdivided here. While a fifth cluster is added to the

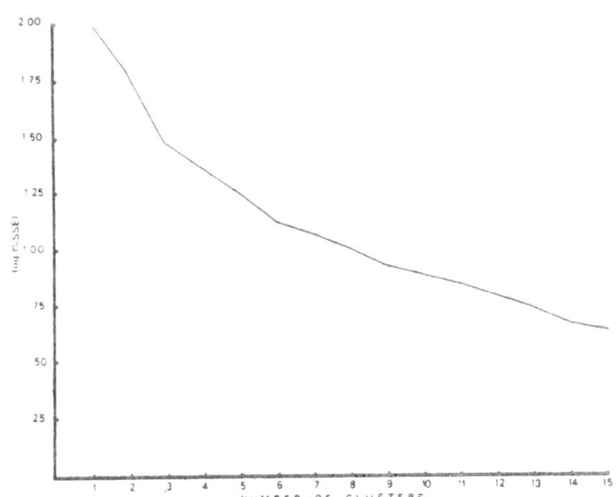

Figure 7. Plot of log(%SSE) Statistics for Caribou Mandibles and Dental Fragments from Pincevent Section 36 Produced by K-means Cluster Analysis to Fifteen Clusters.

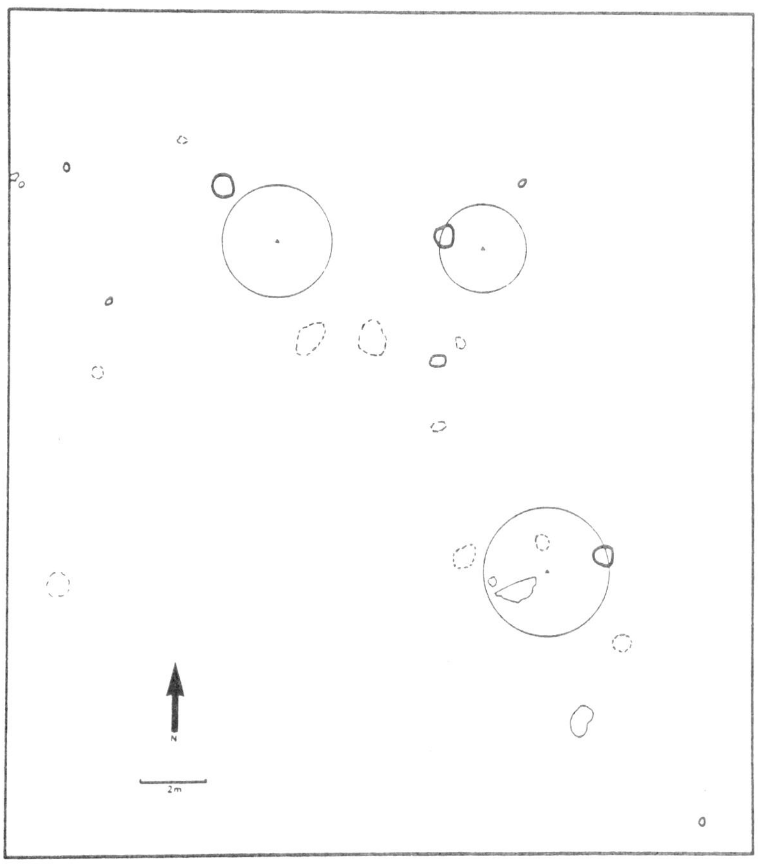

Figure 8. Map of Three Cluster K-means Solution for Pincevent Caribou Mandible Distribution. Dots mark cluster centroids; circles are RMS radii for clusters.

TABLE 2.

Table 2. Three cluster solution statistics for Pincevent mandibular and dental fragments (reproduced from Simek and Larick 1983).

RANK	N	X	Y	RMS
1	132	8.36	6.18	1.51
2	78	14.36	6.36	1.15
3	9	16.00	15.27	1.71

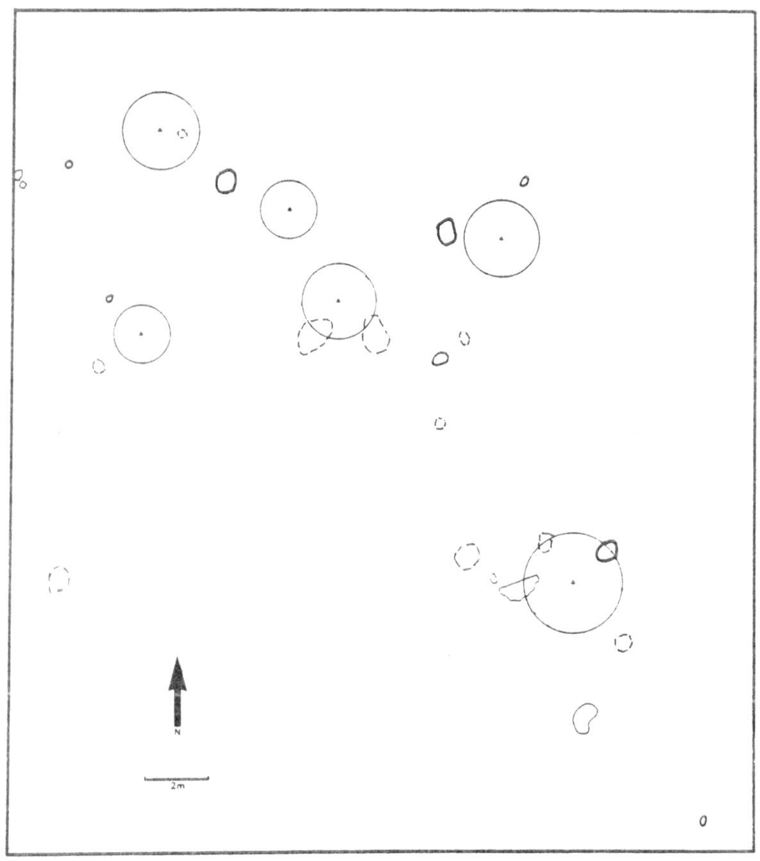

Figure 9. Map of Six Cluster K-means Solution for Pincevent Caribou Mandible Distribution. Dots mark cluster centroids; circles are RMS radii for clusters.

TABLE 3.

Six cluster solution statistics for Pincevent mandibular and dental fragments (reproduced from Simek and Larick 1983).

RANK	N	X	Y	RMS
1	69	14.55	6.18	1.00
2	63	8.73	5.45	0.76
3	59	10.00	8.00	0.98
4	15	5.09	3.45	1.03
5	9	16.73	15.82	1.33
6	4	4.55	8.91	0.76

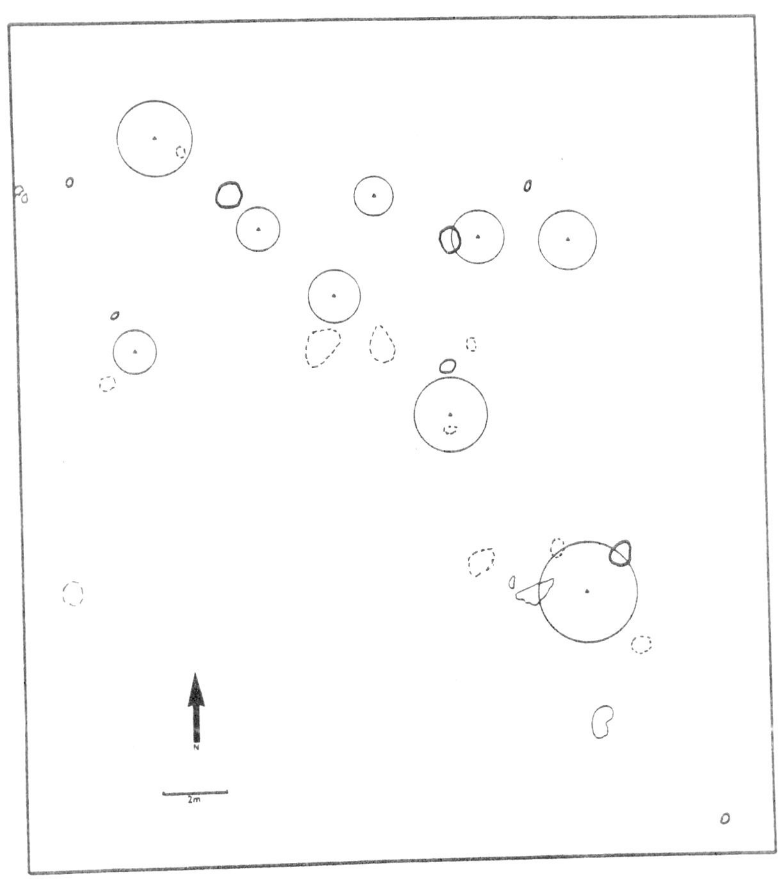

Figure 10. Map of Nine Cluster K-means Solution for Pincevent Caribou Mandible Distribution. Dots mark cluster centroids; circles are RMS radii for clusters.

TABLE 4.

Nine cluster solution statistics for Pincevent mandibular and dental fragments (reproduced from Simek and Larick 1983).

RANK	N	X	Y	RMS
1	56	9.81	7.64	0.71
2	50	13.82	6.00	0.71
3	40	7.82	5.82	0.56
4	23	10.00	4.91	0.51
5	19	16.36	15.82	1.33
6	13	4.91	3.27	1.00
7	9	16.73	15.82	1.33
8	6	13.09	11.09	0.98
9	3	4.18	9.09	0.58

northwest complex, the northeast cluster is divided into three separate clusters, and the southern area is broken into northern and southern units.

Interpreting the patterns defined here requires consideration of cluster point frequency, as this factor seems linked to the spatial patterns analytically defined. In the case of mandibles and teeth, the richest cluster in the first solution is divided into the most clusters initially, and these are further divided as the run proceeds. The second richest cluster is divided only once in the last solution. As was noted above, the clusters having greatest point frequencies seem to reflect, as clustering proceeds, series of smaller concentrations located together. Conversely, clusters having few materials in early solutions are made up of fewer clusters packed into the area. The example of mandibles and teeth should illustrate how the k-means procedure operates to define multiple levels of patterning in the spatial distributions of artifact categories. The examination of other individual classes reveals some interesting patterns in this regard.

Dental materials have their greatest abundance in the northwestern area of Section 36, and most clusters are centered in this area as the clustering process is carried out. Other materials seem to be more evenly dispersed among the three hearth areas. Metatarsals and percoirs have similar patterns of cluster formation through the breakdown process. Figures 11 to 13 illustrate the three, six, and nine cluster solution maps for percoirs. For this artifact class, cluster formation is more uniform in the northern part of the site. At the six cluster level, three clusters are located in the vicinity of the northwest hearth, two are associated with the northeast feature, and a single cluster remains in the southern part of the site. At the nine cluster level, this general uniformity continues. Four clusters occur in the northwest, three in the northeast, and two in the south.

The richest clusters occur closest to the hearths throughout the series of solutions, as was the case for mandibles and teeth. When six clusters are distinguished, 94.1% of all percoirs and 65.5% of all metatarsals are contained in the three clusters closest to the three features. It should be pointed out that these three clusters may not be the richest overall, but two of the three largest point frequency values are located in the northern areas. Nine cluster solutions also place the richest clusters close to the hearths. Fifty percent of all metatarsals occur in the three closest clusters at this level, since two clusters are indicated very near to each of the northern features (Figure 13). When both clusters are considered "closest" to the hearths, a total of five clusters contribute to an 82.4% of all percoirs at the three features. When only the closest clusters are counted for the two northern hearths, 46.3% of all

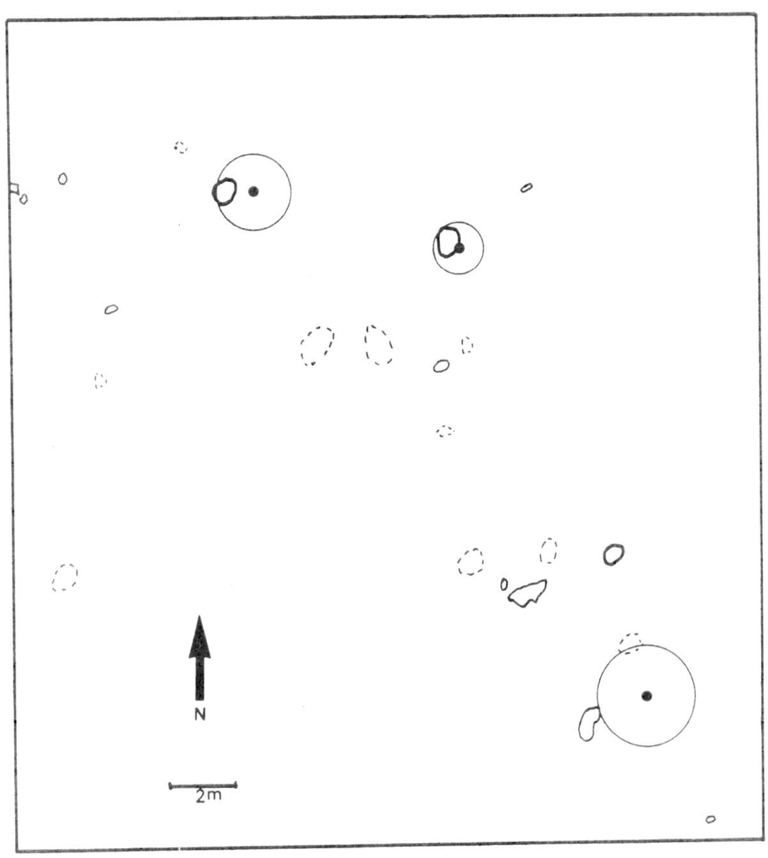

Figure 11. Map of Three Cluster K-means Solution for Pincevent Percoir Distribution. Dots mark cluster centroids; circles are RMS radii for clusters.

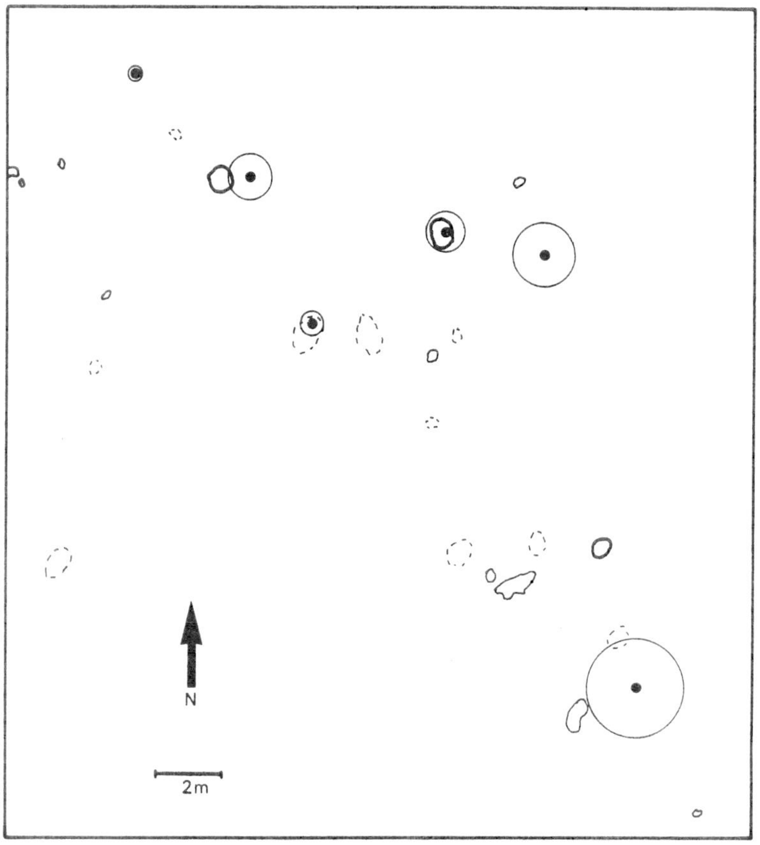

Figure 12. Map of Six Cluster K-means Solution for Pincevent Percoir Distribution. Dots mark cluster centroids; circles are RMS radii for clusters.

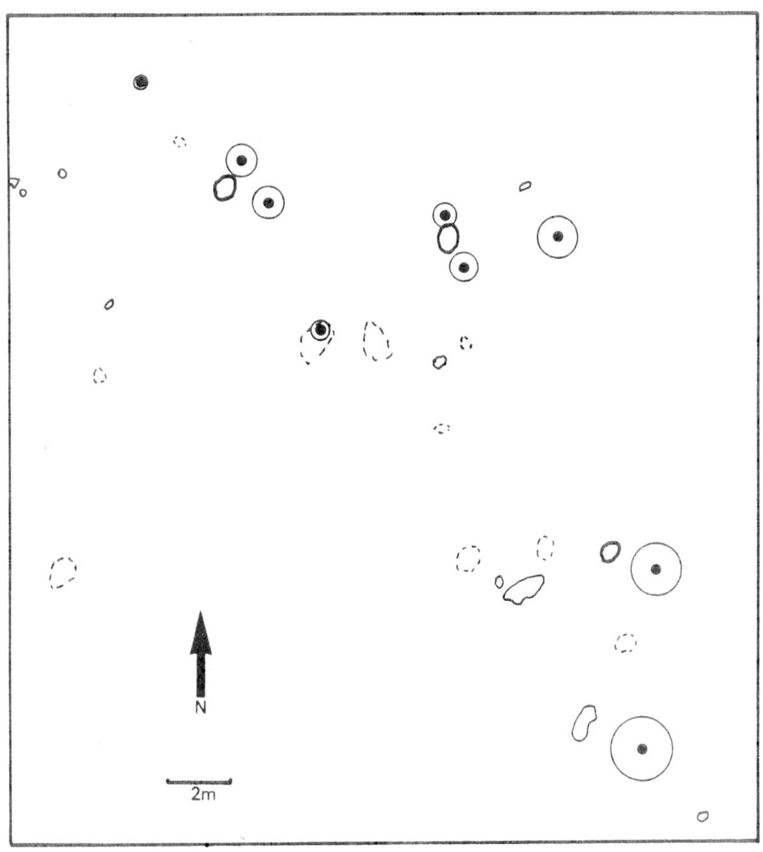

Figure 13. Map of Nine Cluster K-means Solution for Pincevent Percoir Distribution. Dots mark cluster centroids; circles are RMS radii for clusters.

percoirs are located close to the features. Given the proximity of the clusters composing the first percentage, this number is preferred as reflecting percoirs actually associated with the features.

Metacarpals have a similar breakdown between three and six cluster solutions to that discussed above for metatarsals and percoirs. This class of artifact diverges from the other two between the six and nine cluster levels (Figures 14 and 15). The three clusters defined in the northwest area become four in the nine cluster solution. However, in contrast to metatarsals and percoirs, four clusters are also formed in the northeastern area. A single cluster remains near the southernmost hearth. As was the case for percoirs, two clusters are closely associated with each of the two northern features. At the six cluster level, 64.7% of all metacarpals occur in the three clusters closest to the hearth. The same percentage is obtained when five clusters--two each in the northern areas and one in the south--are counted at the nine cluster solution level. When only the three closest clusters are considered at nine clusters, 26.8% of all metacarpals occur in the hearth areas. Both of these figures are lower than percentages obtained for percoirs, suggesting a less compact concentration of metacarpals around the hearths.

Two other material classes will be discussed to illustrate variation in cluster formation. Figures 16, 17, and 18 illustrate maps produced from the three, six, and nine cluster solutions for burins. Here again, the northwest hearth area is divided into three clusters at the six cluster level, and a fourth cluster is added in the nine cluster solution. The other two areas are subdivided in a different sequence than was seen in other artifact classes. From the three cluster solution to the six cluster solution, no division is made in the single cluster associated with the northeastern hearth, while two clusters are formed in the southern area. At the nine cluster level, two clusters occur in the northwest and three are indicated near the southern hearth. This sequence is reversed from that in other artifact groups and suggests a paucity of burins in the northeast hearth area. The lack is surprising since the burin is a classic and always rich component of any Magdalenian industry. It is possible, however, that location rather than abundance conditions these results. If many burins are packed into a small area at the northeastern hearth, then subdivision would not necessarily be related to abundance. In fact, 37.3% of all burins occur in this area and it seems likely that a single compact cluster here reflects a rich local concentration. As is the case generally, burins are concentrated in clusters closest to the hearth features. At the six cluster level, 72.9% occur in the clusters closest to the three hearths; 50.3% occur in the closest clusters at the nine cluster level.

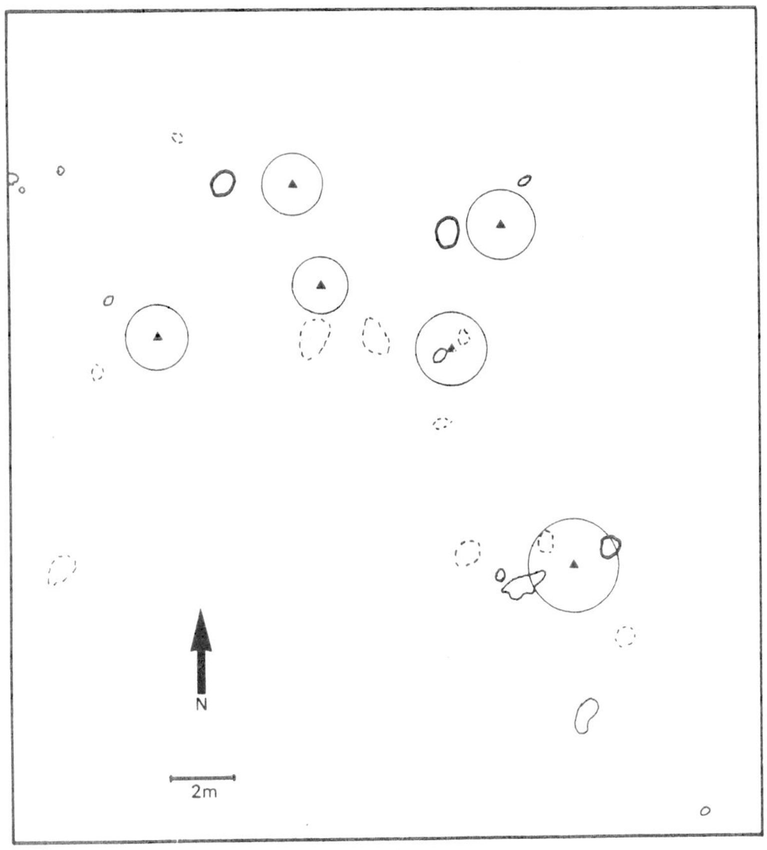

Figure 14. Map of Six Cluster K-means Solution for Pincevent Caribou Metacarpal Distribution. Dots mark cluster centroids; circles are RMS radii for clusters.

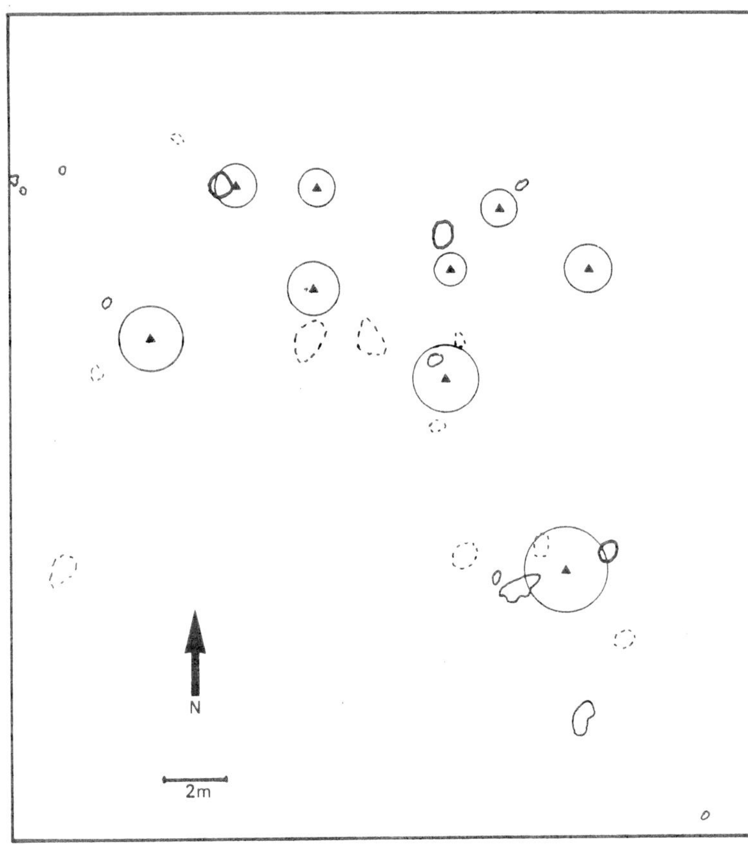

Figure 15. Map of Nine Cluster K-means Solution for Pincevent Caribou Metacarpal Distribution. Dots mark cluster centroids; circles are RMS radii for clusters.

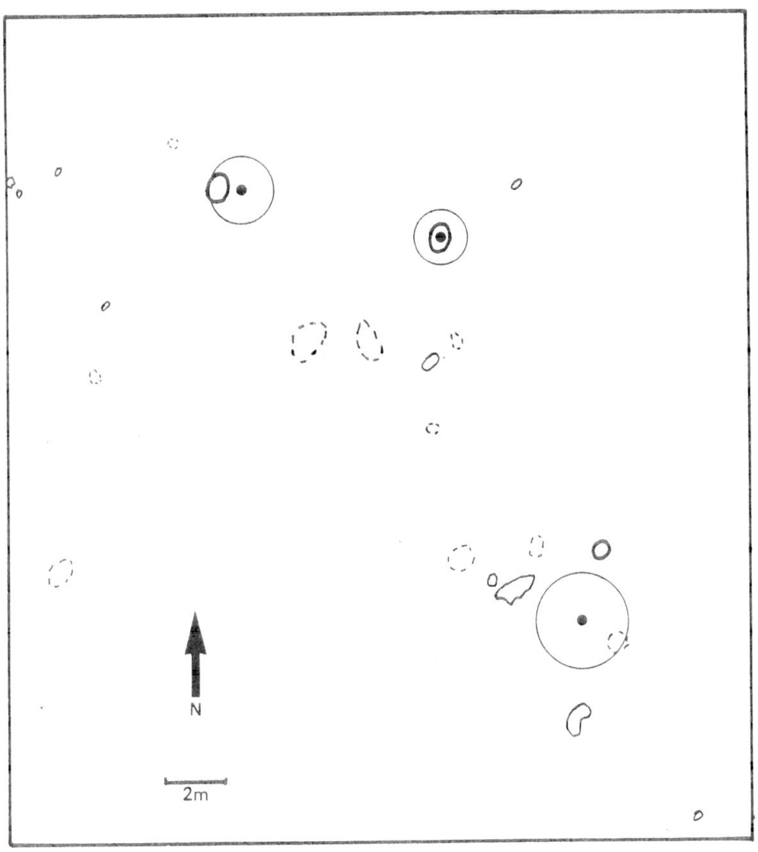

Figure 16. Map of Three Cluster K-means Solution for Pincevent Burin Distribution. Dots mark cluster centroids; circles are RMS radii for clusters.

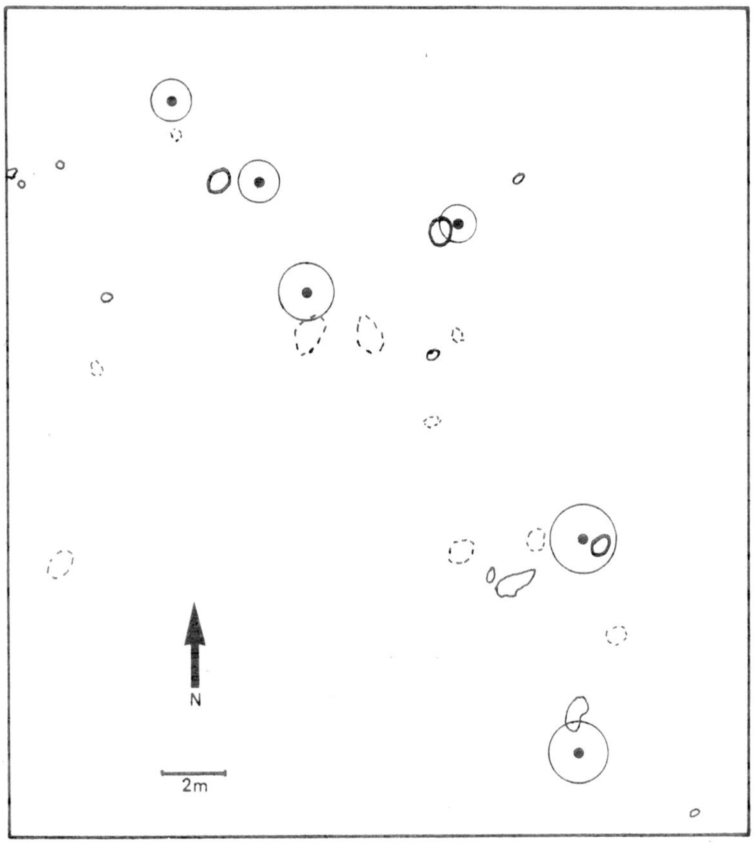

Figure 17. Map of Six Cluster K-means Solution for Pincevent Burin Distribution. Dots mark cluster centroids; circles are RMS radii for clusters.

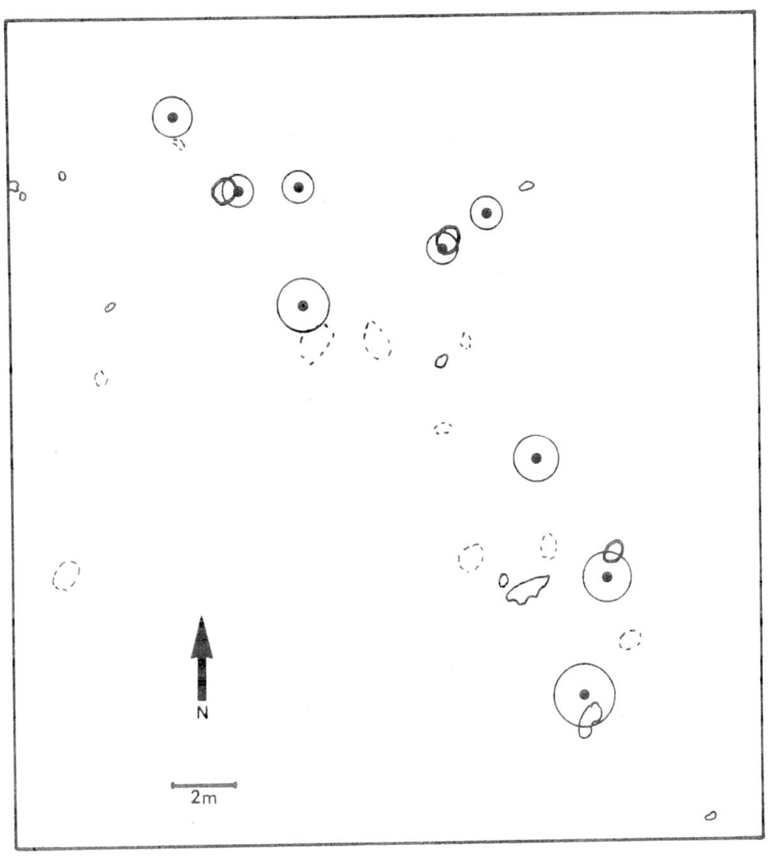

Figure 18. Map of Nine Cluster K-means Solution for Pincevent Burin Distribution. Dots mark cluster centroids; circles are RMS radii for clusters.

Figures 19 and 20 represent the three and six cluster solutions for endscrapers. Divergence from other material classes occurs between these solutions, although cluster patterns produced at the nine cluster level conform to those discussed for burins. At the six cluster level, the northeastern area has three clusters in the hearth area. Only two clusters occur near the northwestern hearth, which normally has the most clusters at this level. A concentration of four endscrapers associated with a small cup-hearth feature in the extreme northeastern sector of the site accounts for this discrepancy. The southern area retains a single cluster characterizing the distributiion of scrapers there. This is divided at the nine cluster level. As is the case for all classes, most of the scrapers occur in clusters closest to the hearths at all clustering levels. Nearly 73% are contained in the three clusters closest to the hearths in the six cluster solutiion. In the nine cluster solution, 55.8% occur in the three closest clusters.

The foregoing discussion of individual analyses is intentionally brief. No attempt has been made to account for all the variation in individual analyses, nor has the subject of interclass associations been raised. This is because variation between classes and associations among classes are relevant issues when contextual evidence is incorporated into the study. Single analyses have been discussed to illustrate the operation of k-means cluster analysis through a series of clustering levels, to indicate the kinds of variation in cluster sequencing that can and do occur, and to examine the effects of artifact abundance on cluster formation. Context and association will be addressed later in this chapter, as will the other issues discussed here, when the concept of cluster zone, defined in Chapter 3, is applied.

The relationship between artifact abundance and location indicated in all of these individual studies deserves some elaboration here. It seems to be a general tendency that more clusters are formed in those areas which have the greatest number of objects within a class. Table 5 shows the raw counts for each artifact category when these are divided into three clusters (see below). The pattern of cluster formation is obviously conditioned by the number of objects present in each of the three initial clusters, but this pattern is best reflected between the three and six cluster levels. In all cases but one, that area with the greatest number of pieces is subdivided first and most. Other areas are divided later in the clustering sequence and/or into fewer clusters. Only percoirs contradict this pattern: the northwest area is divided first and ultimately into the most clusters, even though the northeast hearth area contains more percoirs. This may be due, as was stated above, to the compaction of percoirs into small concentrations near all hearths.

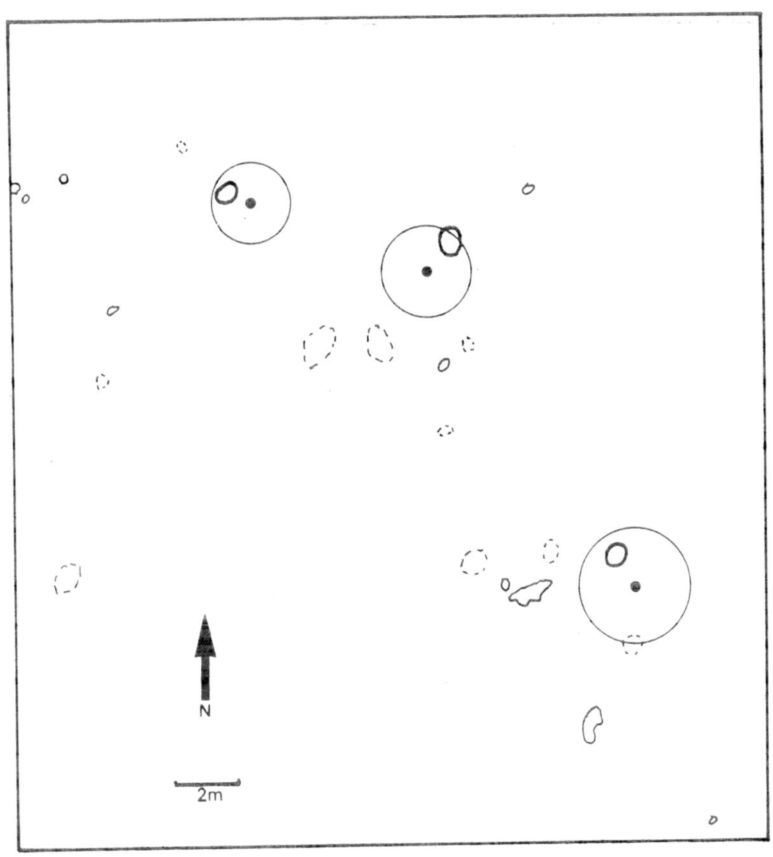

Figure 19. Map of Three Cluster K-means Solution for Pincevent Endscraper Distribution. Dots mark cluster centroids; circles are RMS radii for clusters.

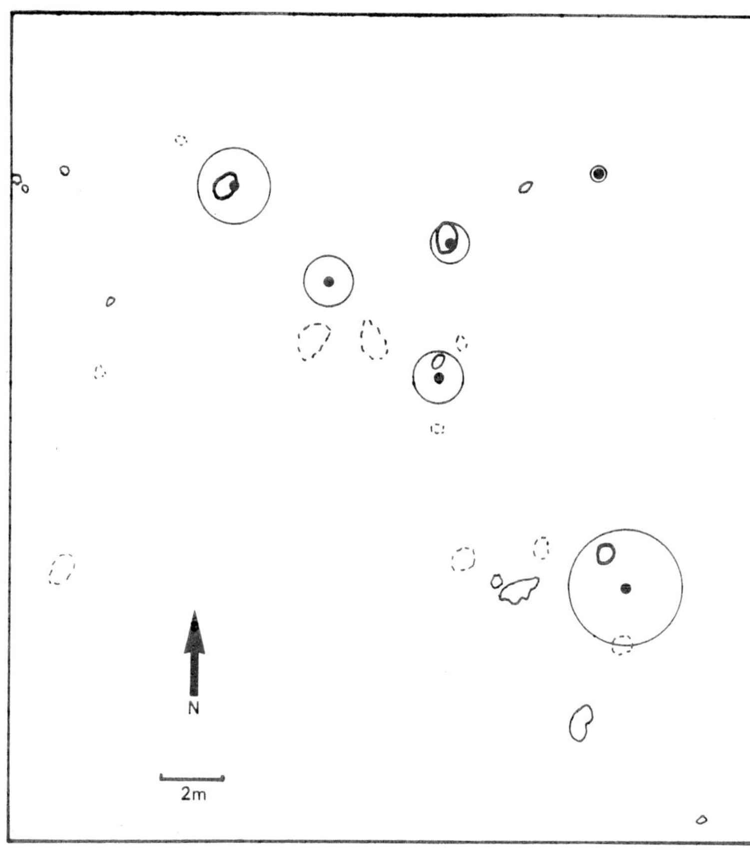

Figure 20. Map of Six Cluster K-means Solution for Pincevent Endscraper Distribution. Dots mark cluster centroids; circles are RMS radii for clusters.

In sum, several factors seem to contribute to spatial variation in materials at Pincevent. These include proximity of clusters to specific types of features, abundance, and degree of artifact concentration. The nature of these factors and how they condition relationships among artifact classes are key points of interest, since the identification of activity specific areas on the site must take them into account. To address the problems of complex deposition and artifact association, however, artifact categories must be examined together in terms of defined analytic--here spatial--units. That is the subject of the second stage of analysis.

STAGE II ANALYSIS: PATTERN ASSOCIATION, ZONE DEFINITION

To display all ten artifact categories together, the ten k-means solution maps produced at each of three clustering levels are overlaid. Three overlay maps are generated, corresponding to three cluster (Figure 21), six cluster (Figure 22), and nine cluster (Figure 23) solutions for all artifact categories. Visual assessment of these overlay maps is more difficult than for individual category maps, but certain patterns are evident.

It should be pointed out first, however, that not all artifact categories showed optimal clustering solutions at these three levels. Figure 24 shows the log(%SSE) curves for all ten artifact classes combined on a single graph. All classes have optimal clustering solutions at the three cluster level. Because the Pincevent analysis was the first test of the k-means method (and we were not entirely certain how the technique would behave at fine cluster scales), we made a decision to use the three cluster solution as a guideline for higher scale maps. Specifically, this decision was made because we were not sure what is indicated by variation in the degree (or strength) of log (%SSE) inflections. Subsequent work suggests that degree of inflection from the early parts of the plot to the higher scale is conditioned by the mathematics of SSE calculation, and that all inflections should be considered useful indications of optimal clustering. However, in this first study, we chose to use multiples of the "strongest" optimal solution scale--three, six, and nine cluster levels--in order to examine multiple patterns. Simek and Larick (1983) discuss this decision in more detail.

All artifact classes show optimal solutions in the neighborhood of six clusters, although the log(%SSE) inflection may be at five clusters or seven clusters in some cases. The same is true for the nine cluster level: some artifact categories may have optimal solutions at the eight cluster level or at ten clusters.

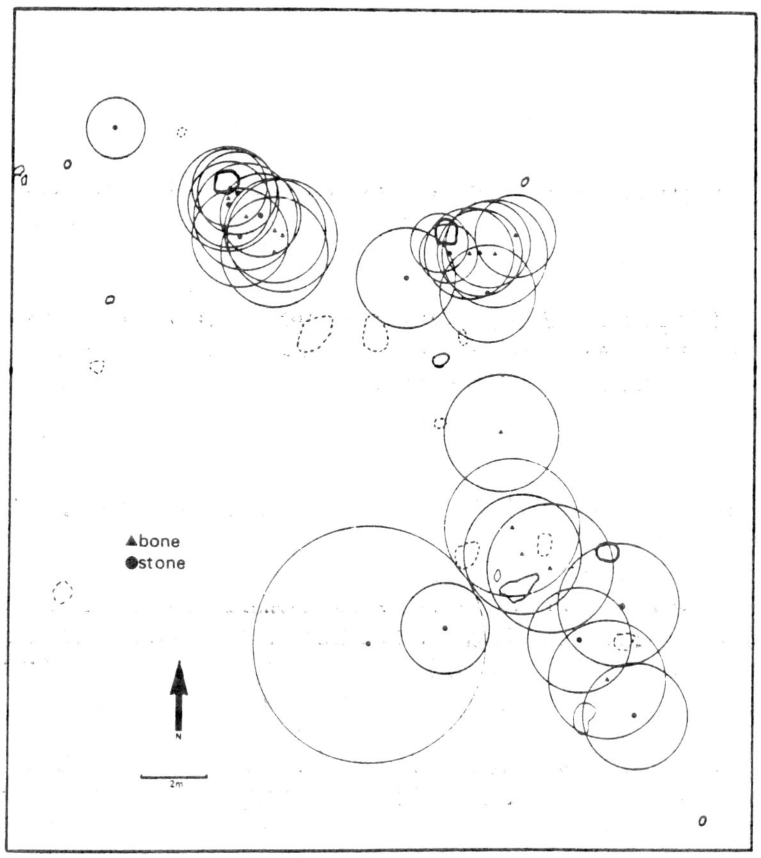

Figure 21. Low Scale Overlay Map Produced by K-means Analyses of Ten Artifact Class Distributions from Pincevent Section 36. Dots mark lithic cluster centroids; triangles mark bone cluster centroids; boxes mark the position of isolated pieces; circles around centroids represent RMS radii for the clusters.

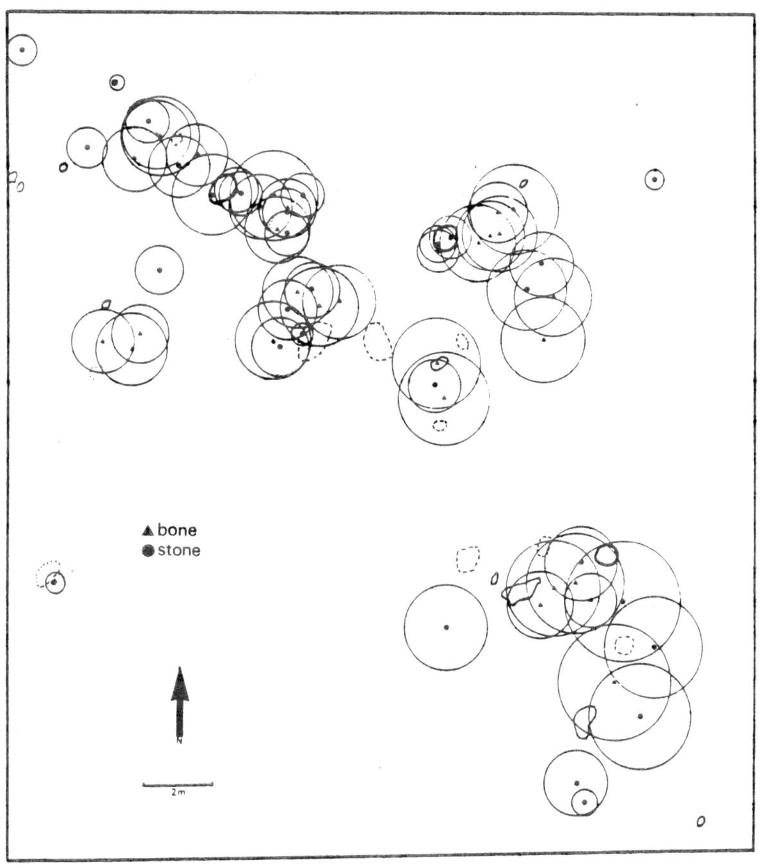

Figure 22. Medium Scale Overlay Map Produced by K-means Analyses of Ten Artifact Class Distributions from Pincevent Section 36. Dots mark lithic cluster centroids; triangles mark bone cluster centroids; boxes mark the position of isolated pieces; circles around centroids represent RMS radii for the clusters.

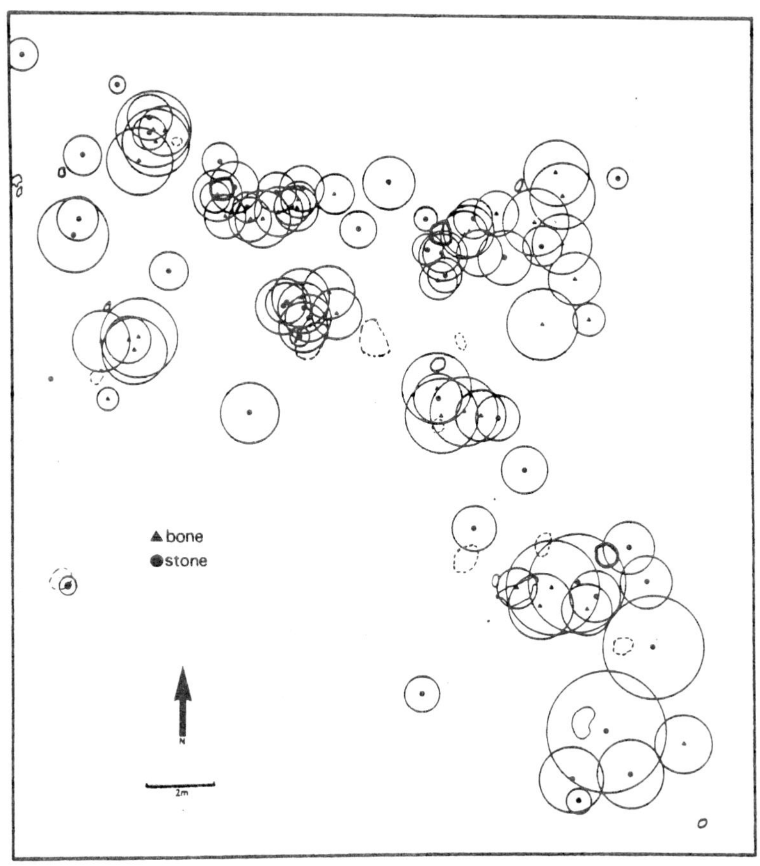

Figure 23. High Scale Overlay Map Produced by K-means Analyses of Ten Artifact Class Distributions from Pincevent Section 36. Dots mark lithic cluster centroids; triangles mark bone cluster centroids; boxes mark the position of isolated pieces; circles around centroids represent RMS radii for the clusters.

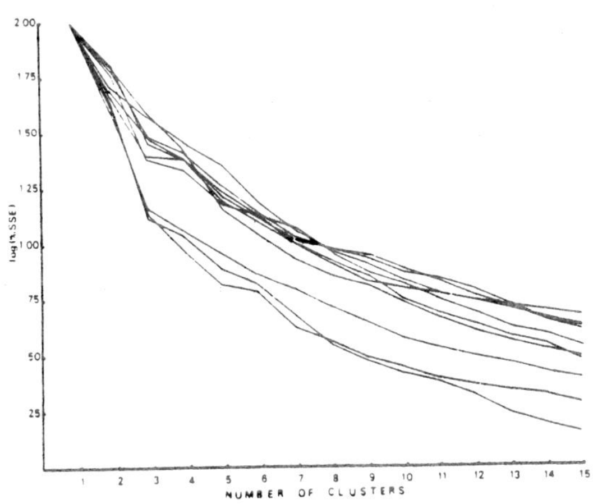

Figure 24. Plot of log(%SSE) Statistics for All Ten Artifact Classes at Pincevent Section 36 as Defined by *K*-means Cluster Analysis to Fifteen Clusters.

The three cluster map presented in Figure 21 is strikingly similar to the interpretations offered by Leroi-Gourhan and Brézillon (see Figure 5 above). Areas of centroid concentration occur near each of the three focal hearths. Two of these, in the northern half of the excavated area, are fairly compact. The third, in the region of the southern hearth, is less so, and clusters here have wider RMS circles. These results compare favorably to the excavators' analyses, defining three general concentrations of deposition that plausibly represent habitation areas. All zones defined by *k*-means analyses are centered on rock-lined features; most centroids are located on one side, that interpreted as "exterior", of those hearths. All three zones contain all artifact categories.

Differences in the individual distributions exist that can be identified at this analytic stage. In all three cluster zones, stone and bone artifact classes are somewhat separate. Stone artifacts tend to be centered more directly on the hearths in the northern zones, while bone objects cluster slightly to the east of the hearths. In the southern area, this separation also exists, although the axis of division runs north and south. Stone objects associated with the southern hearth also seem to be centered on the feature. In all three zones of cluster concentration, hearths serve as the "centers of gravity" for centroids, and stone and bone materials are separate within a general centroid concentration located on one side of the hearth.

The six cluster overlay map (Figure 22) exhibits the same general configuration. A series of clusters is centered on each hearth, with stone and bone centroids still separate to some extent. Another series of cluster zones are formed at this level, however, and these tend to be centered on some of the exterior features. Clusters making up these zones are less rich than those near focal features, and many are more dispersed since they have larger RMS circles associated with the centroids. Stone tool clusters are more compact than bone, and their centroids tend to occur in closer proximity to each other. In contrast to stone, bone materials have more dispersed clusters and these are located further from the hearth features. Small cluster zones defined at the six cluster level show variation in content, with some made up exclusively of bone and others of stone tools.

Figure 23, representing the nine cluster overlay map, not only illustrates the trends already noted but also displays some new characteristics. Cluster zones centered on the hearths in earlier solutions are subdivided here. The separation of bone from stone clusters is maintained, with lithic centroids located close to features and bone further to the "exterior". Other clusters occur near peripheral features, particularly in the southern part of the site. In this area, two principal zones of deposition can now be distinguished: a larger, more general one near the hearth itself and a second, more dispersed, centered on a gravel feature. A gravel-and-stain feature located between the two northern hearth areas shows a complex concentration of all stone tool classes together with a few bone categories. Another cluster zone to the east contains primarily bone artifact classes. Other peripheral features are the centers of gravity for smaller cluster assemblages.

The compactness of clusters centered on the three principal hearths suggests that artifact clusters at these loci are denser and more restrained spatially than those located near the peripheral features. A link is again indicated between the location of clusters in relation to certain contextual features and the relative abundance of materials. This relationship is noted through several levels of resolution, i.e., in several characterizations of the artifact distributions, although it is less clear at the nine cluster level than in simpler configurations. This is probably due to a refinement of the description to include smaller and less compact cluster zones.

The maps discussed above can now serve to form zones or spatial units for further analyses. In the Pincevent material distributions, it should be noted that zones are, for the most part, completely separated in space. In a few cases, a very slight RMS overlap may occur between zones, but this overlap is small when compared to that within zones.

For the three cluster solutions, zone definition is easy. Three areas of centroid concentration and RMS overlap coincide with the focal hearths. Accumulations occur to a single side of the hearth, and centroids tend to be compact in the northern two cluster zones and more dispersed in the south. Figure 25 shows the three zones defined at this level.

Zone definition based on the six cluster overlay map is somewhat more difficult, although here, too, there are clear separations between areas of centroid concentration and RMS overlap. Six zones are evident (Figure 26), while five isolated clusters are scattered away from the primary areas of concentration. Slight overlap can be seen between zone 1, associated with the northwest focal hearth, and zone 3 to the south. The overlap involves a single cluster in each zone, and occurs at the extreme edge of both RMS circles. Given the heavy RMS overlap reflected within both zones, a zone boundary seems reasonable here. In the southern area, zone 6 has been defined to include a pair of clusters which, while overlapping, are slightly separate from the rest of the zone. The dispersed nature of all materials in the southern area warrants a conservative attitude towards zone definition. Therefore, these slightly distinct clusters are lumped with the larger centroid/RMS mass.

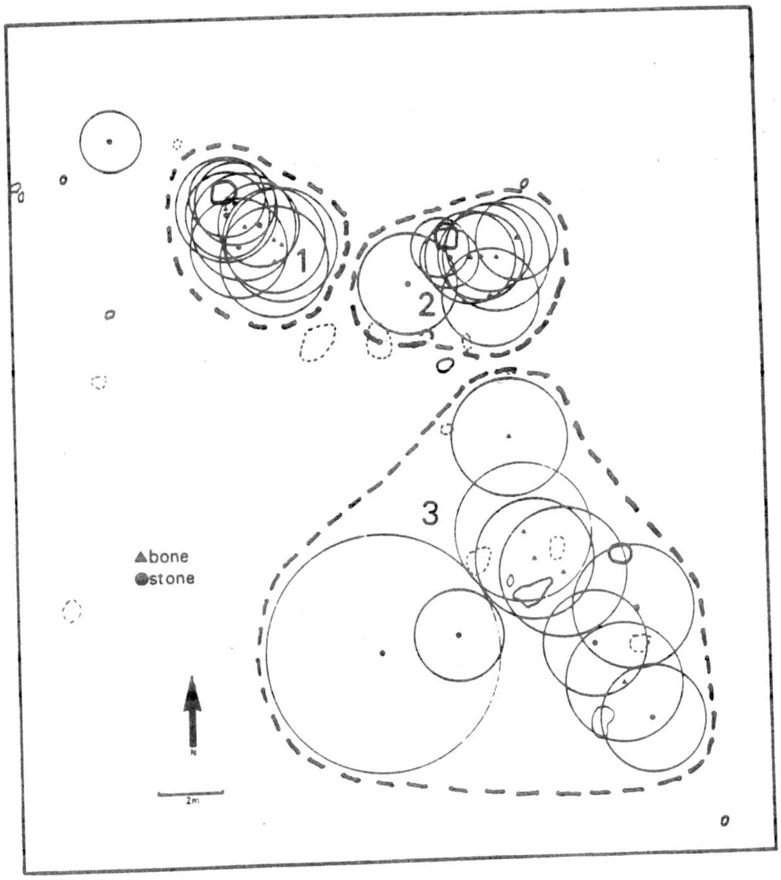

Figure 25. Three Zones Defined at Low Scale Solution for Ten Artifact Class Distributions from Pincevent Section 36. Zones are outlined with a dotted line.

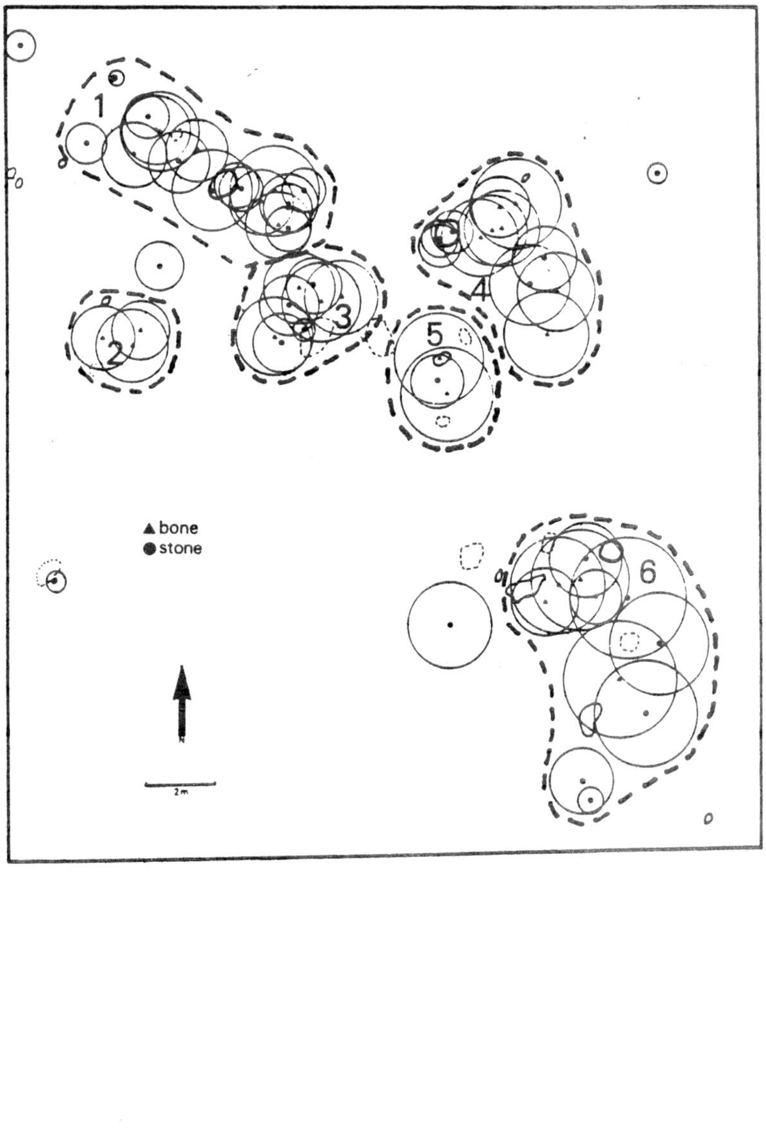

Figure 26. Six Zones Defined at Medium Scale Solution for Ten Artifact Class Distributions from Pincevent Section 36. Zones are outlined with a dotted line.

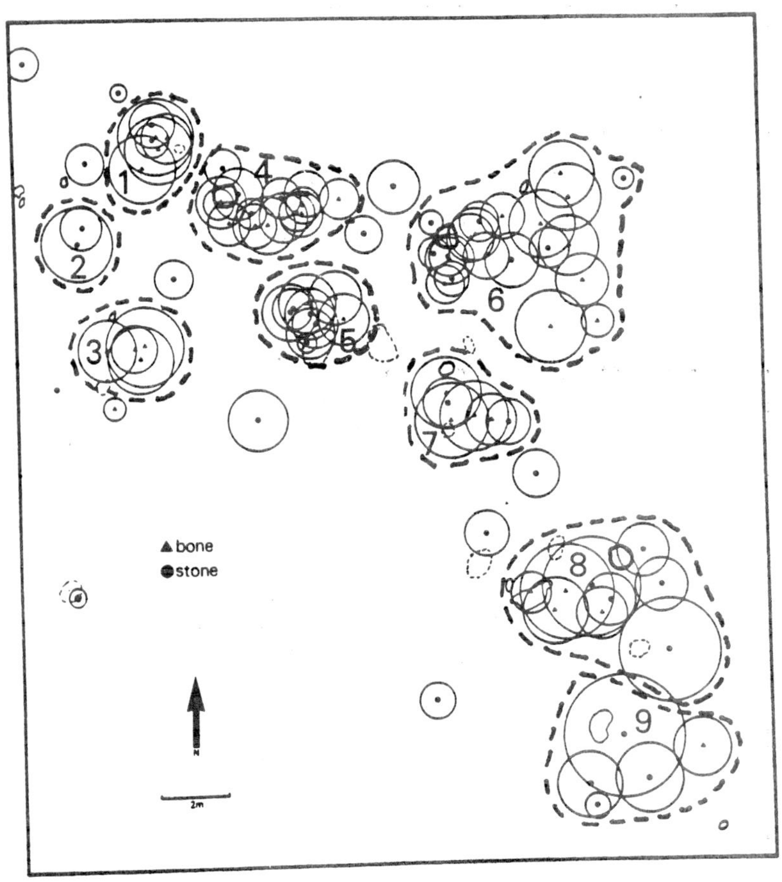

Figure 27. Nine Zones Defined at High Scale Solution for Ten Artifact Class Distributions from Pincevent Section 36. Zones are outlined with a dotted line.

However, another cluster is isolated from this zone. Other zones distinguished at this level are well-differentiated and need no detailed justification.

Nine zones are identified based on the nine cluster solution overlay map (Figure 27). In addition, seven isolated clusters are present. Zones 1 and 3 both include nearby stone tool clusters which show no RMS overlap with other clusters in the zones. These are very near the centroid masses in both cases. Three northern zones all have very compact centroids, and some individual clusters nearby can be assigned to particular zones with little difficulty. All small clusters in the north contain few artifacts, a characteristic they do not share with other isolated clusters (which tend to be rather heavy concentrations of a single artifact class). In the southern sector, two zones are defined, although a slight overlap is present in single cluster RMS circles from each zone.

A final note on this analytic stage is important. The number of cluster zones defined at each solution level coincides with the number of clusters plotted onto the maps for each artifact class (with the exception of isolated clusters). This is a rather surprising result, since there is no *a priori* reason to expect this kind of spatial conformity. That there is a basic similarity among all data classes when each has been analyzed separately, and that this similarity should be expressed with an equal number of zones as individual clusters for each level, is *not* preconditioned by the technique. Thus, zones represent areas where all, or most, artifact classes concentrate into discrete areas. However, the number of areas distinguished at each level is not determined by the number of clusters of each class plotted onto the map.

STAGE III ANALYSIS: PATTERN ASSOCIATION, CONTENT ANALYSIS

The relationship between artifact abundance and cluster location may best be examined using artifact frequencies within cluster zones. However, raw counts may give a distorted picture of the relations among classes, since those categories having high frequencies in the site as a whole (e.g., backed bladelets) may overwhelm less numerous classes. As was outlined in Chapter 3, two kinds of data standardization will be used here to assess content and abundance variation among the spatial units defined as zones. First, point frequencies are replaced with class rank values by zone. Second, raw counts are converted to (1) class maximum, and (2) zone maximum percentages.

Assessing Global Association among Artifact Classes

Tables 5, 6, and 7 present the raw counts and corresponding class rank values for zones at the low, medium, and high scales. As an initial assessment of the relation between rank (i.e., relative abundance) and location, the rank of each artifact class cluster is plotted at the centroid of that cluster. Figure 28 shows the rank map for the three cluster solution (i.e., low scale). Other rank maps will be discussed shortly.

With few exceptions (these occurring in the northern part of the site), ranks for clusters distribute sequentially into the three zones. In general, the highest ranks are located in the northwest area near hearth V105, second ranks are located near hearth T112, and lowest ranks occur near hearth L115. All artifact classes are present in all zones, and within a zone the rank values for all class frequencies are similar.

The general pattern noted at the three cluster level--like ranks concentrated in space--holds in the six and nine cluster solutions. At the six cluster level (Figure 29), clear patterning is still evident. High-ranking clusters (1 and 2) tend to occur in groups, as do middle ranking ones (3 and 4). Low ranking clusters are more scattered around the excavated area, but they generally occur away from the focal hearths. Although more difficult to evaluate visually, the nine cluster rank map (Figure 30) exhibits the same relationships on a more detailed scale. The highest ranking clusters are concentrated near the two northern hearth features, where they tend to occur in small groups. The southern hearth area, least rich of the three focal hearths, contains mostly middle rank clusters. Finally, a series of peripheral zones are identified at the nine cluster solution. Located away from the central hearths, these contain a few low ranking clusters, are scattered across the excavated area, and tend to center on smaller peripheral features.

Use of this rank mapping procedure allows the recognition of a recurring relation between location and abundance at Pincevent. In general, the high ranking clusters occur together, i.e., cluster centroids for most of the major artifact categories show similar high rank values when found in spatial proximity. This is the case regardless of the total amount of materials found at a given location. Importantly, this pattern is observed at each of the three clustering levels examined for this study. Middle ranking clusters show a similar pattern, but low ranking clusters (6 to 9) are more scattered, even isolated in space, at the nine cluster level.

To examine relationships among classes based on these ranks, Spearman's rho statistic for rank order correlation was calculated among classes at each spatial scale. Table

TABLE 5.

Raw counts and ranks for ten artifact classes from Pincevent 36 based on zones defined at the lowest clustering solution scale. Rank values are in parentheses.

MATERIAL CLASS	CLUSTER ZONE 1	2	3
Metacarpals	109 (1)	105 (2)	21 (3)
Metatarsals	89 (1)	85 (2)	12 (3)
Mandibles	134 (1)	78 (2)	16 (3)
Antlers	75 (2)	89 (1)	0 (3)
Ribs	232 (1)	104 (2)	30 (3)
Percoirs	44 (2)	82 (1)	10 (3)
Endscrapers	40 (2)	79 (1)	13 (3)
Burins	87 (1)	66 (2)	24 (3)
Cores	94 (1)	38 (2)	17 (3)
Bladelets	254 (1)	0 (3)	72 (2)

TABLE 6.

Raw counts and ranks for ten artifact classes from Pincevent 36 based on zones defined at the medium clustering solution scale.

MATERIAL CLASS	CLUSTER ZONE 1	2	3	4	5	6
			COUNTS			
Metacarpals	49	9	51	87	23	16
Metatarsals	68	0	21	85	0	12
Mandibles	78	8	59	69	0	14
Antlers	64	0	28	70	0	2
Ribs	225	14	0	87	23	17
Percoirs	39	0	7	80	0	10
Endscrapers	32	0	34	40	10	13
Burins	73	0	21	58	0	25
Cores	69	0	32	32	0	9
Bladelets	491	0	0	0	0	72
			RANKS			
Metacarpals	3	6	2	1	4	5
Metatarsals	2	7.5	3	1	7.5	4
Mandibles	1	5	3	2	7.5	4
Antlers	2	7.5	3	1	7.5	4
Ribs	1	5	7.5	2	3	4
Percoirs	2	7.5	4	1	7.5	3
Endscrapers	3	7.5	2	1	5	4
Burins	1	7.5	4	2	7.5	3
Cores	1	7.5	2.5	2.5	7.5	4
Bladelets	1	7.5	7.5	7.5	7.5	2

TABLE 7.

Raw counts and rank values for ten artifact classes from Pincevent 36 based on zones defined at the highest clustering solution scale.

MATERIAL CLASS	CLUSTER ZONE 1	2	3	4	5	6	7	8	9
				COUNTS					
Metacarpals	0	0	9	50	49	9	16	16	0
Metatarsals	19	0	3	0	19	64	0	5	7
Mandibles	13	0	7	63	56	32	6	14	0
Antlers	21	0	4	42	24	56	10	2	0
Ribs	37	0	14	156	35	64	19	17	0
Percoirs	4	0	0	0	7	10	0	3	7
Endscrapers	5	0	0	42	26	8	8	9	4
Burins	19	0	0	55	19	19	4	13	8
Cores	21	4	0	25	29	28	0	9	0
Bladelets	291	32	0	166	0	0	0	0	2
					RANKS				
Metacarpals	8.5	8.5	5.5	1	2	5.5	3.5	3.5	8.5
Metatarsals	2.5	8.5	6	8.5	2.5	1	8.5	5	4
Mandibles	5	8.5	6	1	2	3	7	4	8.5
Antlers	4	8.5	6	2	3	1	5	7	8.5
Ribs	3	8.5	7	1	4	2	5	6	8.5
Percoirs	4	8.5	8.5	8.5	2.5	1	8.5	5	2.5
Endscrapers	6	8.5	8.5	1	2	4.5	4.5	3	7
Burins	2.5	8.5	8.5	1	2.5	2.5	6	4	5
Cores	4	6	8.5	3	1	2	8.5	5	8.5
Bladelets	1	3	8.5	2	8.5	8.5	8.5	8.5	4

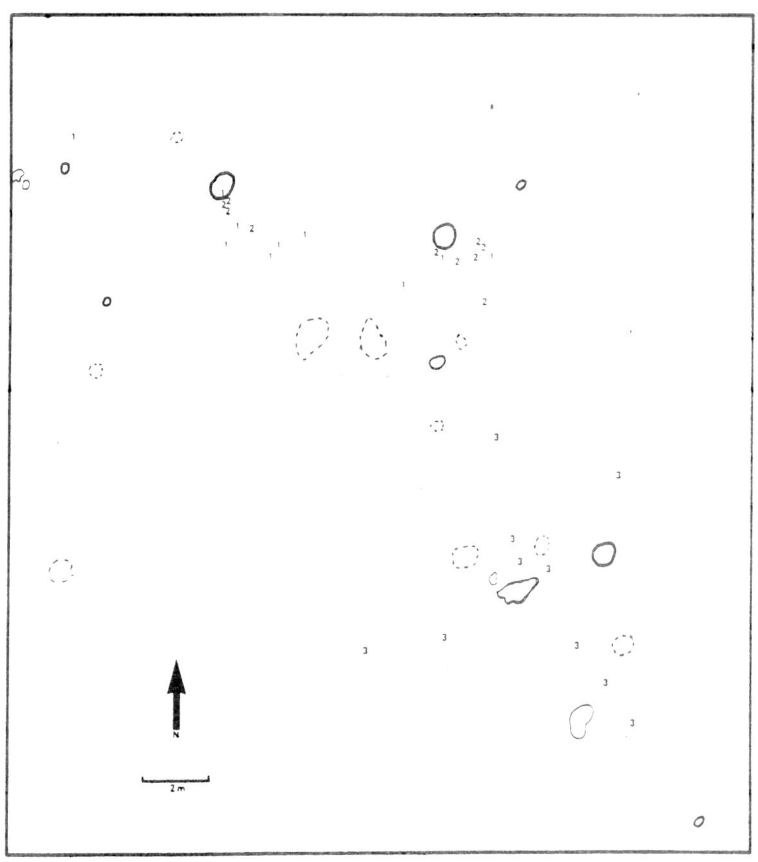

Figure 28. Map of Cluster Point Frequency Ranks Plotted at Centroids for Three Cluster Solutions at Pincevent Section 36.

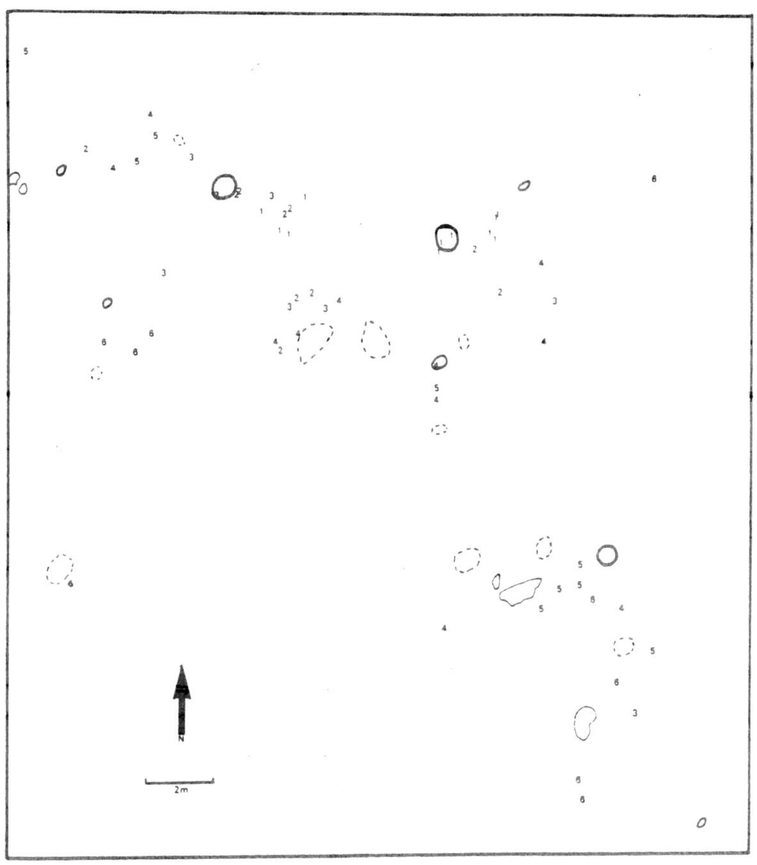

Figure 29. Map of Cluster Point Frequency Ranks Plotted at Centroids for Six Cluster Solutions at Pincevent Section 36.

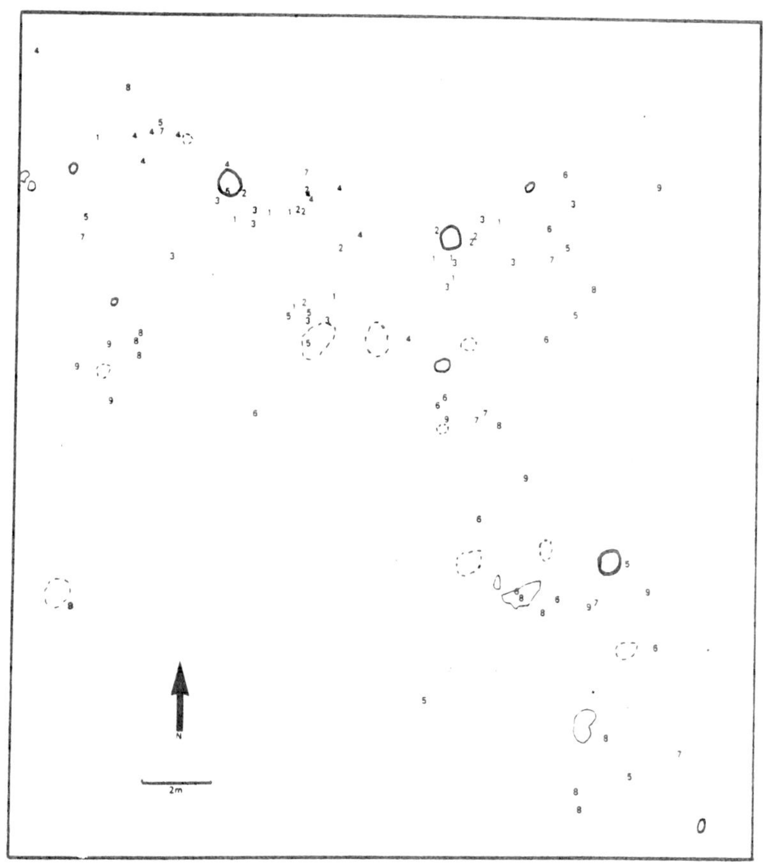

Figure 30. Map of Cluster Point Frequency Ranks Plotted at Centroids for Nine Cluster Solutions at Pincevent Section 36.

8 presents the <u>rho</u> matrix for the low scale solution (three zones) but is not very informative since only three variables--the zones--are observed for each artifact class. To be significant in this table, a correlation coefficient must attain unity in either positive or negative direction.

Tendencies noted on the plotted rank maps are evident here again. Most artifact classes have their most abundant clusters in the northwest hearth area (Zone 1). Zone 2 at the northeast hearth contains mostly second ranking clusters. The southeastern hearth area is associated only with third ranking clusters. Therefore, all artifact classes are linked through correlations with the exception of backed bladelets.

Table 9 shows the correlation matrix resulting from the six cluster zone divisions (see Table 6 for zone rankings). Here, a somewhat more complex picture emerges. Two data classes are independent from all others at this level. Caribou ribs have no significant correlation with any other artifact category, and bladelets have four significant negative correlations but no other significant links. Metacarpals and endscrapers are correlated, and these have (rather weak) links to metatarsals and antler fragments through the endscraper class. All other data classes are intercorrelated at high levels of statistical significance. These results suggest a similar picture to that present at the three cluster zone level, i.e., general interrelation among all but a few artifact classes over space. Thus, it would seem that all classes are distributed in similar fashion.

Table 10 shows the rank order correlation matrix for nine cluster zones. Again, backed bladelets have no strong positive links to other data classes. Metacarpals have significant correlations with endscrapers--as was the case for six zones--and with metatarsals. Percoirs are linked to this group through a significant positive correlation with metatarsals and have no correlation with other artifact categories. A second group of artifact classes is linked to the first through endscrapers. This includes mandibles, antlers, ribs, burins, cores (as well as endscrapers).

Three sets of artifact classes may be indicated by this last correlation matrix, although they are interrelated. One set involves a pair of categories, metacarpals and endscrapers. A second set also contains two classes, percoirs and metatarsals. The final set contains all other artifact categories except bladelets and is linked to the first group through scrapers. Endscrapers and metacarpals are also correlated at the six-zone level, while all other classes were linked with the exceptions of ribs and bladelets.

TABLE 8.

Matrix of Spearman's rank order correlation statistics (rho) for all Pincevent artifact classes when three cluster zones are considered. Artifact classes are numbered as follows: (1) metacarpals, (2) metatarsals, (3) mandibles, (4) antlers, (5) ribs, (6) percoirs, (7) endscrapers, (8) burins, (9) cores, and (10) backed bladelets. All correlations of 1.0 have (p < .01).

	1	2	3	4	5	6	7	8	9	10
1	/									
2	1.0	/								
3	1.0	1.0	/							
4	0.5	0.5	0.5	/						
5	1.0	1.0	1.0	0.5	/					
6	0.5	0.5	0.5	1.0	0.5	/				
7	0.5	0.5	0.5	1.0	0.5	1.0	/			
8	1.0	1.0	0.5	0.5	1.0	0.5	0.5	/		
9	1.0	1.0	0.5	0.5	1.0	0.5	0.5	1.0	/	
10	0.5	0.5	0.5	-0.5	-0.5	-0.5	-0.5	0.5	0.5	/

TABLE 9.

Matrix of Spearman's rank order correlation statistics (*rho*) for all Pincevent artifact classes when six cluster zones are considered. Artifact classes are numbered as follows: (1) metacarpals, (2) metatarsals, (3) mandibles, (4) antlers, (5) ribs, (6) percoirs, (7) endscrapers, (8) burins, (9) cores, and (10) backed bladelets. *Rho* values with (p < .05) are starred; those with (p < .01) are underlined.

	1	2	3	4	5	6	7	8	9
1	/								
2	.50	/							
3	.39	.94*	/						
4	.50	1.00	.94*	/					
5	-.09	-.39	-.33	-.39	/				
6	.33	.94*	.89*	.94*	-.39	/			
7	.88*	.76	.65	.76	-.30	.65	/		
8	.21	.89*	.94*	.89*	-.14	.94*	.54	/	
9	.37	.90*	.99*	.90*	-.48	.81	.64	.90*	/
10	-1.00	-.93*	-.56	-.93*	-.74	-.61	-1.00	-.24	-.54

TABLE 10.

Matrix of Spearman's rank order correlation statistics (rho) for all Pincevent artifact classes when nine cluster zones are considered. Artifact classes are numbered as follows: (1) metacarpals, (2) metatarsals, (3) mandibles, (4) antlers, (5) ribs, (6) percoirs, (7) endscrapers, (8) burins, (9) cores, and (10) backed bladelets. Rho values with (p < .05) are starred.

	1	2	3	4	5	6	7	8	9
1	/								
2	-.34	/							
3	-.74*	.24	/						
4	.52	.32	.83*	/					
5	.52	.21	.85*	.95*	/				
6	-.46	.91*	.11*	.16	-.07	/			
7	.83*	.03	.84*	.64	.73*	-.06	/		
8	.39	.37	.78*	.71*	.81*	.38	.80*	/	
9	.33	.43	.81*	.71*	.70*	.37	.64	.71*	/
10	-.82*	-.52	-.41	-.39	-.16	-.57	-.41	-.11	-.25

When the results of correlation analyses are related to the *k*-means cluster maps presented earlier, the class groups noted here can be given spatial relevance. The first two artifact groupings seem to relate to differences in hearth area content in the northern sector of the site. The nine-zone rank matrix (Table 7) shows that endscrapers and metacarpals both have their highest frequencies in zones 3 and 4. These zones correspond to the northwestern hearth area and to the zone between the northeast and northwestern hearth.

The second class grouping--metatarsals and percoirs--has high rank values in zone 5 (the northeast hearth area). These differences suggest variability in the activities carried out at the two loci, although variation in bone content may reflect social factors such as meat distribution, division of labor, or composition of the consuming group. These two class groupings are linked through one or several of their members to other groups, including the most general third grouping of artifact classes; this suggests either particular artifact forms were used in a variety of ways or that intervening factors influenced artifact location between tool use and abandonment.

Pincevent artifact distributions may reflect the operation of systematic disposal behaviours. As was discussed earlier, complexity of cluster zone content should remain the same as smaller spatial subdivisions are made if dumping occurred. This assumes, of course, that larger accumulation areas are made up by multiple disposal events. The mixing of activity specific refuse into dump areas produces its own spatial variation in zone content, and homogeneity results. Smaller zones are reflections of less intense dumping, perhaps due to less intense use of the loci, but still represent smaller versions of larger accumulation piles.

To examine change in homogeneity as smaller spatial units are distinguished, each correlation table is characterized by the proportion of the total table taken up by statistically significant positive correlations. This value is plotted against the spatial scales obtained by *k*-means multi-pattern recognition, and a profile of homogeneity is generated (Figure 31). The plot obtained from the Pincevent analyses shows a more or less straight line across all levels of spatial refinement. This suggests a disposal mode of pattern generation at Pincevent, and conforms to the expectations outlined earlier (Chapter 2) for such a set of behaviours. Differences among spatial zones do exist, and these may reflect some functional variation across the site. However, when examined in detail, these always seem to relate to differences between hearth areas, i.e., the three zones identified at the lowest spatial scale.

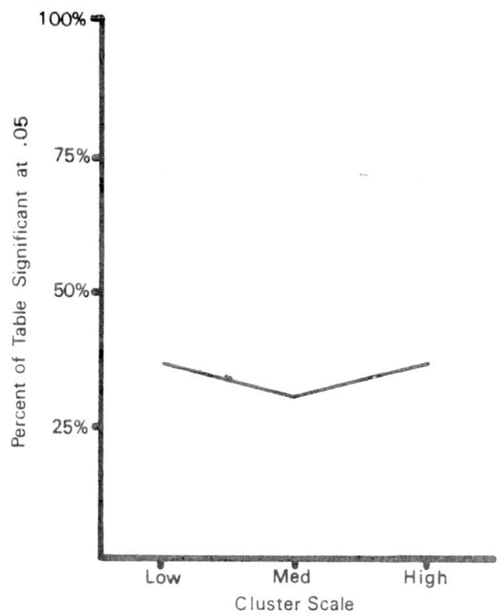

Figure 31. Plot of Accumulation Content Homogeneity with Scale of Spatial Refinement for Pincevent Section 36.

Assessing Local Associations among Artifact Classes

Vectors representing zone content standardized by class and zone maximum values were computed for all spatial units at each scale, according to the technique outlined in Chapter 3. The vectors, referred to here as "transformations", will be discussed by spatial scale below. The lowest, three zone, scale is considered first, followed by the six zone and nine zone levels.

Table 11 presents the values obtained after transformation within classes for the three cluster zones defined in k-means analysis. It is clear that one zone, corresponding to hearth area V105, contains the greatest number of objects in most classes. Those not having their highest value in zone 1--antlers, percoirs, and endscrapers--nevertheless have relatively high percentages of the maximum in this zone. Zone 2, centered at hearth T112, also shows high values for all material classes except for bladelets. Classes not having a value of 100, or maximum, in zone 1 have that value in this second zone. The third zone shows all values lower than one-third of the overall maximum for each class. Antlers are absent in the third zone, while burins and bladelets show the highest values occurring within this zone.

The information in this table corresponds closely to that obtained from rank matrices at the three cluster level. V105 is the richest of the three zones, T112 is the second richest, and L115 the poorest in terms of material abundance. However, certain similarities and differences can be seen that were not evident in the rank matrix. Percoirs and endscrapers, for example, have similar values in all three zones. Antler fragments, having the same rank pattern, still show a different proportional distribution than scrapers and percoirs. Other groups are noted: metacarpals and metatarsal are similarly distributed; ribs, mandibles, and cores have the same distributions of values. Other classes have unique distributions of standardized values over the three zones.

Figure 32 shows the circle graphs for class-maximum transformations. Each graph in the figure corresponds to a spatial zone at the three cluster level. Clearly, zone 1 shows the highest percentages of all materials, both stone and bone, except for percoirs and endscrapers. Zone 2 shows relatively high values for most classes, although ribs, cores, and bladelets are not well represented here. Cluster zone 3 shows small values for all materials, although burins and bladlets have slight peaks.

The transformation by class is designed to control for variation between artifact categories in terms of their overall abundance at the site. Therefore, if a strict location/abundance factor were operating, i.e., if deposition resulted in the placement of larger or smaller dumps of equivalent content, then a series of three

TABLE 11.

Matrix of class maximum standardization values for Pincevent 36 artifact classes when three cluster zones are considered. All values are rounded to the nearest integer.

MATERIAL CLASS	CLUSTER ZONE 1	2	3
Metacarpals	100	96	19
Metatarsals	100	96	13
Mandibles	100	58	12
Antler	84	100	0
Ribs	100	45	13
Percoirs	54	100	12
Endscrapers	51	100	16
Burins	100	76	28
Cores	100	40	18
Bladelets	100	0	28

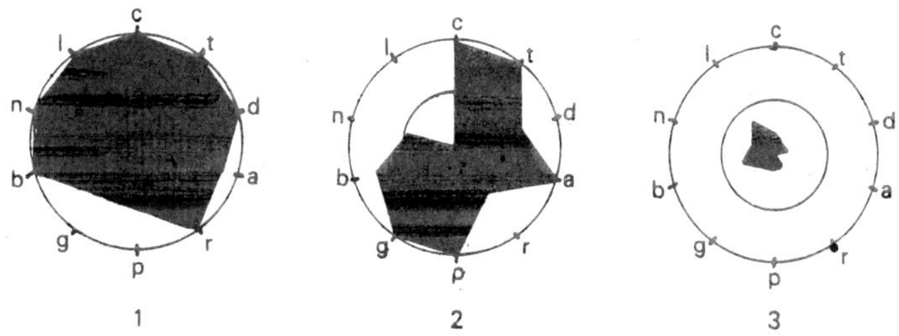

Figure 32. Circular Graphs for Class Maximum Standardization of Zone Contents when Three Zones Are Considered for Pincevent Section 36. Radii correspond to artifact classes as follows: (c) metacarpals, (t) metatarsals, (d) mandibles, (a) antler, (r) ribs, (p) percoirs, (g) endscrapers, (b) burins, (n) cores, (l) bladelets. Circumference of circle represents class maximum value of 100.

different sized, roughly circular polygons would be expected. This is the case for zones 1 and 3, where nearly all artifact classes are represented in all three zones. However, zone 2 contains burins, endscrapers, antlers, percoirs, metacarpals, and metatarsals in abundance, while other classes are present in reduced numbers. It seems that zones 1 and 3 are equivalent in terms of their constituent elements while zone 2 is slightly different.

Table 12 shows the transformed values for the Pincevent three zone solution when percentages are computed by zone. Figure 33 shows the circle graphs corresponding to this table. Here, the similarity of zones 1 and 3 is evident, and can be attributed to high relative percentages of bladelets and caribou ribs. Smaller peaks occur for burins and mandibles in both graphs. The only difference between the two zones seems to be in the antler present. Zone 2 shows a high bone content, in contrast to the two other zones. Coupled with this relative abundance of all bone classes, stone tool classes poorly represented in zones 1 and 3 form a high percentage of the assemblage in zone 2.

In comparing the results of the two transformations, two dimensions of variation can be defined. First, as illustrated by the within class percentages, there is a pattern of artifact abundance relating to location. The richest zone overall, zone 1, contains the richest clusters for most object categories. Zone 2 contains high percentages of many classes but reduced numbers of other classes. Zone 3, the poorest zone in terms of absolute abundance, contains roughly equivalent percentages of most artifact classes, and seems to represent a less rich version of cluster zone 1. Second, the relative contribution of various artifact classes to each zone accentuates this pattern. Again, zones 1 and 3 are similar in content, with both having bladelets and ribs as dominant elements of the zone assemblages. Zone 2 differs from the other two in its high bone content and its high percentages for burins, scrapers, percoirs and cores.

Percent-of-maximum transformations were performed on raw count tables for the Pincevent six zone solution. Table 13 shows the values produced when data are transformed by class maximum, and Table 14 presents the results of transformation by zone maximum. Figures 34 and 35 display these results graphically.

Three kinds of zones seem to be present at this cluster level when examining class maximum transformations. Zone 1, centered on hearth V105, shows high values for all classes but percoirs and metacarpals. Zones 3, 4, and 6 contain few or no bladelets and lack ribs as well. Zones 2 and 5 are small zones containing low percentages of a few artifact categories. Of particular interest here is the similarity in graphs shown

TABLE 12.

Matrix of zone maximum standardization values for Pincevent 36 artifact classes when three cluster zones are considered. All values are rounded to the nearest integer.

MATERIAL CLASS	CLUSTER ZONE 1	2	3
Metacarpals	43	100	29
Metatarsals	35	81	17
Mandibles	53	74	22
Antler	30	85	0
Ribs	91	99	42
Percoirs	17	78	14
Endscrapers	16	75	18
Burins	34	63	33
Cores	37	36	27
Bladelets	100	0	100

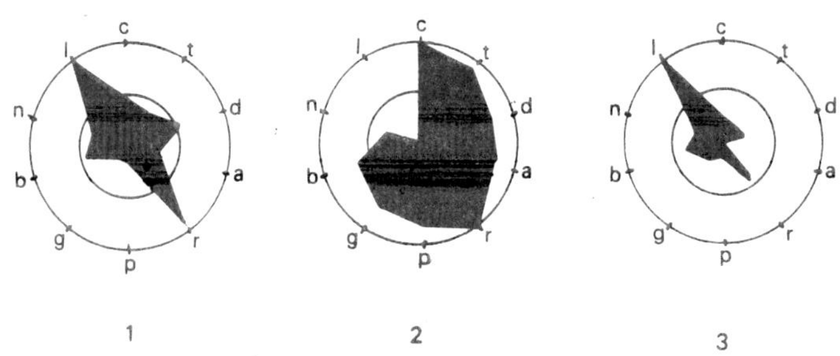

Figure 33. Circular Graphs for Zone Maximum Standardization of Zone Contents when Three Zones Are Considered for Pincevent Section 36. Radii correspond to artifact classes as follows: (c) metacarpals, (t) metatarsals, (d) mandibles, (a) antler, (r) ribs, (p) percoirs, (g) endscrapers, (b) burins, (n) cores, (l) bladelets. Circumference of circle represents class maximum value of 100.

TABLE 13.

Matrix of class maximum standardization values for Pincevent 36 artifact classes when six cluster zones are considered. All values are rounded to the nearest integer.

MATERIAL CLASS	CLUSTER ZONE 1	2	3	4	5	6
Metacarpals	56	10	58	100	26	18
Metatarsals	80	0	25	100	0	14
Mandibles	100	10	76	88	0	17
Antler	91	0	40	100	0	3
br Ribs	100	6	0	39	10	8
Percoirs	49	0	9	100	0	13
Endscrapers	80	0	85	100	25	33
Burins	100	0	29	79	0	34
Cores	100	0	46	46	0	13
Bladelets	100	0	0	0	0	15

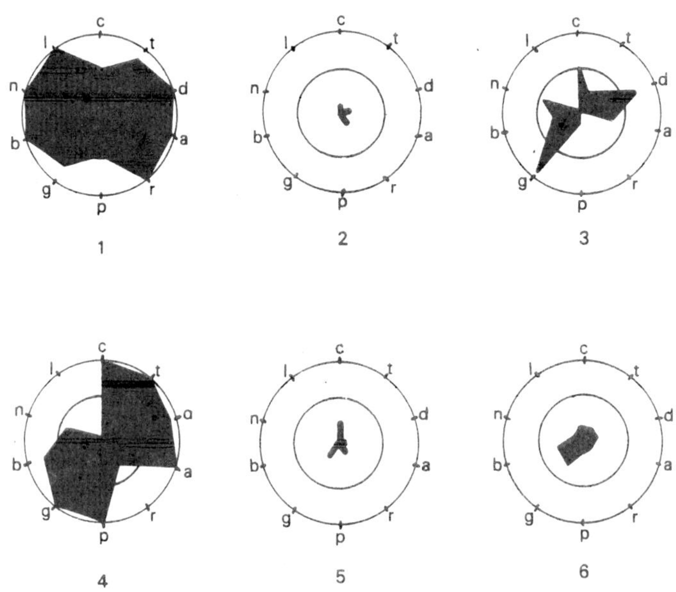

Figure 34. Circular Graphs for Class Maximum Standardization of Zone Contents when Six Zones Are Considered for Pincevent Section 36. Radii correspond to artifact classes as follows: (c) metacarpals, (t) metatarsals, (d) mandibles, (a) antler, (r) ribs, (p) percoirs, (g) endscrapers, (b) burins, (n) cores, (l) bladelets. Circumference of circle represents class maximum value of 100.

TABLE 14.

Matrix of zone maximum standardization values for Pincevent 36 artifact classes when six cluster zones are considered. All values are rounded to the nearest integer.

MATERIAL CLASS	CLUSTER ZONE 1	2	3	4	5	6
Metacarpals	10	64	86	100	100	22
Metatarsals	14	0	36	98	0	17
Mandibles	16	57	100	79	0	19
Antler	13	0	47	80	0	3
Ribs	46	100	0	100	100	24
Percoirs	8	0	12	92	0	14
Endscrapers	7	0	58	46	43	18
Burins	15	0	36	67	0	35
Cores	14	0	54	37	0	13
Bladelets	100	0	0	0	0	100

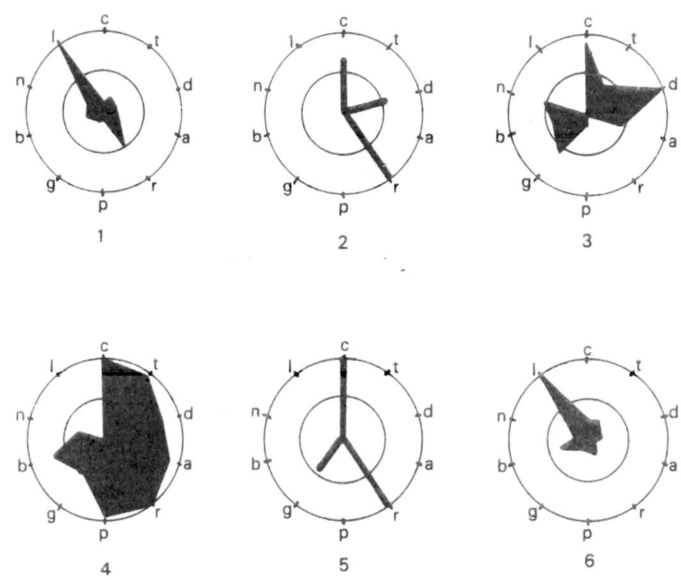

Figure 35. Circular Graphs for Zone Maximum Standardization of Zone Contents when Six Zones Are Considered for Pincevent Section 36. Radii correspond to artifact classes as follows: (c) metacarpals, (t) metatarsals, (d) mandibles, (a) antler, (r) ribs, (p) percoirs, (g) endscrapers, (b) burins, (n) cores, (l) bladelets. Circumference of circle represents class maximum value of 100.

by Zones 3, 4, and 6. Polygons are of varying size, but the general aspect of all three graphs is the same. Peaks occur for metacarpals, mandibles, endscrapers, and cores. Zone 3 contains few burins, but zones 4 and 6 show peaks for that class. Zone 4 has a greater value for percoirs than the other zones, while zone 6 is somewhat lacking in antler fragments when compared to the other two zones. In all, these three cluster zones are very similar given the differences in overall material abundance.

The locations of these three similar units provide interesting points of comparison. Zones 4 and 6 are hearth-centered, while zone 3 is the area between hearths V105 and T112 interpreted as a communal dump by the excavators. At the three zone level, it was the hearth-centered zones at V105 and L115 that were similar, but two factors contribute to the shift seen here. First, a large number of backed bladelets, associated with L115 at the three zone solution, are now spatially isolated. The extraction of these bladelets from the L115 assemblage brings it into line with zone 4 at T112. The separation of zone 3 (the "dump") seems to have altered the profile of the zone centered on V105 so that it no longer resembles zone 6. If zone 3 is a dump, then it is not surprising that the two hearth zones are similar, since a generalized deposition of all materials can be expected there under the disposal behaviour hypothesis. The divergence of zone 1, on the other hand, is not expected. However, the major difference between this dense zone and the other three is in the high percentages of bladelets and ribs present. In other respects, it has a similar graph to the hearth-centered zones and to the dense concentration between zones 1 and 4.

Different sets of cluster zones are suggested at the medium solution scale by zone maximum transformations. Bladelets and ribs dominate zones 1 and 6, repeating the pattern seen at the three zone level. This is somewhat surprising because a substantial number of bladelets were isolated from the zone centered on L115 between clustering levels. Bladelets seem to be concentrated into small areas in large numbers, and when one of these concentrations is isolated, as in this case, other concentrations within the same zone provide sufficient numbers of bladelets to maintain numerical dominance.

Two small zones, 2 and 5, have high percentages of metacarpals and ribs. Zone 2 also contains mandible fragments, while zone 5 contains a few endscrapers. These zones are low density units, yet the fact that they both are dominated by the same two bone categories suggests some similarity.

The final two zones, 3 and 4, are distinct from the other pairs in content. They have points of similarity between them but exhibit important differences. Both have comparable percentages of metacarpals, mandibles, antlers, scrapers, burins and cores. Zone 4, centered on T112, also shows high relative percentages for metatarsals, ribs, and percoirs. The latter zone again shows a rather general aspect, containing all artifact classes except bladelets.

At the six cluster level, a total of five clusters are isolated from cluster zones. Three of these isolated clusters contain bladelets in large numbers. One cluster contains seven cores and is located next to a small hearth feature in the southwestern part of the excavated area. In the extreme northeast, a pocket of three endscrapers is centered on another small cup hearth. Of particular importance is the tendency for bladelets to occur away from zone areas. Half of the bladelet clusters produced by k-means analysis are isolated at this clustering level, and these contain 38.49% of all objects in the class. This supports the observation made earlier, that bladelets are concentrated in small areas. Here, however, it is evident that some of these small but dense concentrations are located away from the main deposition locations.

To summarize the observations made on the basis of percent-of-maximum transformations at the six zone solution, two dimensions of unit similarity are again indicated. For class maximum transformations, three zones (3, 4, and 6) have similar profiles. Two of these zones are centered on hearth features, while the third is a dense area between two hearths previously interpreted as a dump. Zone 1 is somewhat different than the previous three on this measurement scale. Also centered on a hearth feature, zone 1 contains equivalent numbers of all artifact categories except metacarpals and percoirs. It seems to be a more generalized version of the "dense" zone pattern. Two sparse zones contain few materials, but these are of similar character. Both small zones contain metacarpals and ribs. Zone 2 has mandibles associated with the other two bone classes, while zone 5 has a few scrapers included in the assemblage. One dimension of variation, then, entails the density of materials located in given zones. Similar profiles are present based on the class maximum transformation in zones having large numbers of objects overall. Even in small, less dense deposit areas, zone content is similar.

The second dimension of variation is based on zone maximum transformations, and highlights the role of bladelets and ribs as dominant artifact classes when they are present in a zone. As has been discussed, bladelets seem to concentrate in small pockets. These pockets contain large numbers of pieces, and they become numerically important whenever they occur in a zone. However, bladelets have a somewhat different distribution than other tool and bone categories and are frequently found outside of dense zones associated with habitation features. Ribs do not seem to be concentrated in small pockets. When ribs are present in a zone, there is a single cluster, and this contains many ribs. The numerical dominance of this bone class may be related to two factors. Ribs are easily broken into many pieces, and they therefore provide more points on the map than do less easily fragmented bone classes. Ribs are also the most numerous single bone in a vertebrate skeleton. In areas of heavy pedestrian traffic, e.g., at the opening of a structure, a dense but dispersed concentration of rib fragments might be anticipated. The Pincevent six zone solution seems to display this pattern, supporting the excavators' interpretations of the hearths. Other characteristics of zone content illustrated by zone maximum transformations conform to those discussed under class maximum transformation.

Table 15 and Figure 36 show class maximum transformations performed on the Pincevent nine cluster solution counts by zone. Again, a basic distinction is apparent between zones made up of a few classes (2, 3, and 9) and richer zones where most, if not all, categories are represented. A single zone lies between these extremes, zone 7, that contains six of the ten artifact groups. As was the case at the six zone level, zones centered on hearths T112 and L115 show similar assemblage profiles to the dense zone between the two northern features. Again, these zones contain few bladelets and rib fragments, and are dominated by a rather consistent blend of metacarpals, metatarsals, mandibles, antlers, percoirs, scrapers, burins, and cores. These zones (5, 6, and 8) reflect general deposits of most artifact categories. Zones 1 and 4 are similar to the first group, with certain important differences. Zone 1, to the northwest of hearth V105, has a much higher bladelet component than does the generalized group; in other respects it is the same. Zone 4 has an elevated bladelet count, but this tool class does not dominate here. Rib fragments, on the other hand, are well represented, having their greatest frequency in this zone. A final difference between zone 4 and the zones described above as "general" is the lack of metatarsals. However, except in zone 6, this bone type is not present in great frequency in any of the aforementioned zones. All of the richest zones, then, seem to represent generalized deposits of most artifact classes, although some classes are more abundant in one or another spatial unit. The two hearth areas T112 and L115 are most similar in relative artifact content. The third hearth area, V105, has abundant bladelets associated with it, but it is similar in profile to the above zones in other respects. Zone 5 between the northern hearths is similar to a hearth area in content, and so is a new zone (1) centered on a small cup hearth to the northwest of V105.

The smaller zones defined at this clustering level are made up of a reduced set of categories. Zone 2 contains a pair of lithic classes, cores and bladelets, but only a few of each are present. Zone 3 contains only bone, and mirrors zone 2 at the six zone level, with

TABLE 15.

Matrix of class maximum standardization values for Pincevent 36 artifact classes when nine cluster zones are considered. All values are rounded to the nearest integer.

MATERIAL CLASS	CLUSTER ZONE 1	2	3	4	5	6	7	8	9
Metacarpals	0	4	18	100	98	18	32	32	0
Metatarsals	30	0	5	0	30	100	0	8	11
Mandibles	21	0	11	100	89	51	10	22	0
Antlers	38	0	7	75	43	100	18	4	0
Ribs	24	0	9	100	22	41	12	11	0
Percoirs	40	0	0	0	70	100	0	30	70
Endscrapers	12	0	0	100	62	19	19	21	10
Burins	35	0	0	100	35	35	7	24	15
Cores	54	10	0	64	100	74	0	23	0
Bladelets	100	11	0	57	0	0	0	0	1

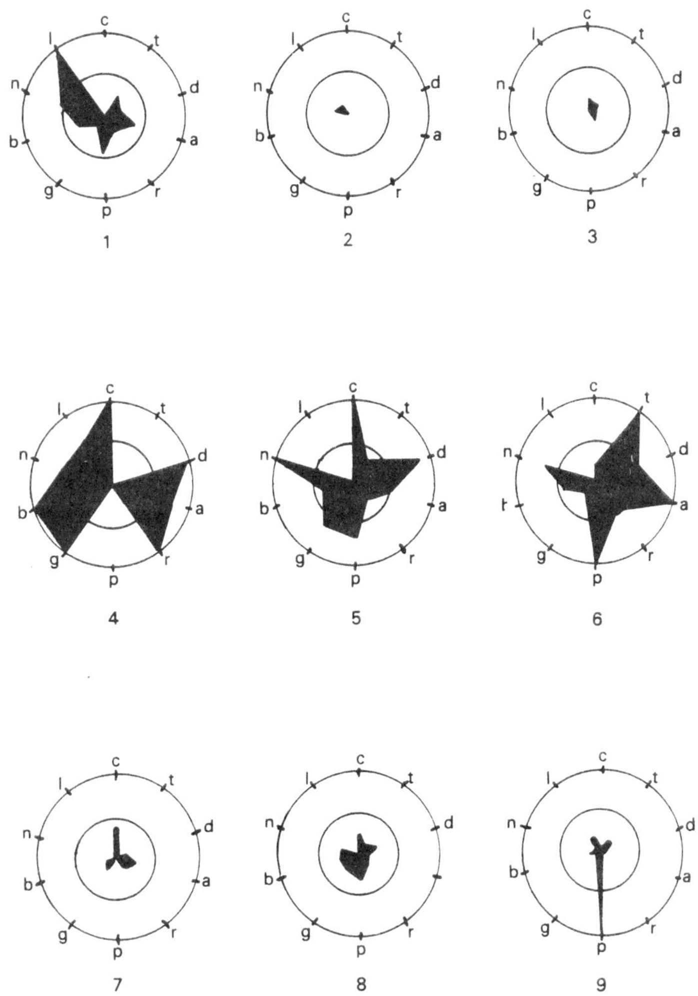

Figure 36. Circular Graphs for Class Maximum Standardization of Zone Contents when Nine Zones Are Considered for Pincevent Section 36. Radii correspond to artifact classes as follows: (c) metacarpals, (t) metatarsals, (d) mandibles, (a) antler, (r) ribs, (p) percoirs, (g) endscrapers, (b) burins, (n) cores, (l) bladelets. Circumference of circle represents class maximum value of 100.

metatarsals and antlers added to the assemblage. Zone 7 contains more artifact classes than other small units, but no class is really abundant. Bone classes are dominant, but also associated are a few burins and scrapers. Zone 9 is made up primarily of lithics, with metatarsals the only bone class represented. Cores are not present in zone 9.

Zone maximum transformations are presented in Table 16 and illustrated in Figure 37. The rich zones defined on the basis of class maximum transformations (1, 4, 5, 6, and 8) are of two kinds here. One pair, zones 1 and 4, have a bladelet component that dominates the assemblage; the presence and importance of ribs in zone 4 is also clear using this transformation. The other three zones can still be characterized as general. Zones 5 and 8 are most similar in content, with metacarpals, mandibles and ribs well represented. Zone 6 contains relatively few metacarpals and scrapers and shows a frequency of rib fragments more like zones 1 and 4 than other general zones.

Sparse zone 2 is clearly dominated by bladelets, although there are a few cores present, and these make up a higher within-class proportion than do the bladelets. Another sparse unit dominated by lithic categories is zone 9. Here, burins and percoirs are important elements in the assemblage, while metatarsals are also present in large numbers. Two zones dominated by bone elements are 3 and 7. Zone 3 is made up entirely of bone materials, with ribs the best-represented class. Zone 7 has a similar profile to 3 but lacks metatarsals; two lithic classes, scrapers and burins, are included in the zones.

For the nine zone solution, seven isolated clusters are distributed across the site in areas away from the major hearth-centered zones. An interesting pattern emerges at this level; only bladelets and cores are isolated by the _k_-means analyses. Bladelets are isolated in four clusters, containing 33.57% of all objects in this class. Cores are found three times in isolated clusters, accounting for 15.33% of the total. It is interesting to note that one of the sparse cluster zones, number 2, contains only cores and bladelets in association. But there is no pattern of spatial relation evident in the scattered isolates. Two isolated clusters, both bladelet concentrations, occur in the northwestern part of the excavated area to the west of hearth V105. A cluster of seven cores is centered on a small hearth in the southwest part of the site. Two bladelet clusters are found near L115 to the west of the feature, but this distribution cannot be related to the clusters west of V105 since Leroi-Gourhan and Brezillon place the inferred structure openings with opposite orientations. A core cluster is located to the south of V105 in the most open part of the excavated area, and another is found between T112 and L115.

TABLE 16.

Matrix of zone maximum standardization values for Pincevent 36 artifact classes when nine cluster zones are considered. All values are rounded to the nearest integer.

MATERIAL CLASS	CLUSTER ZONE 1	2	3	4	5	6	7	8	9
Metacarpals	0	0	64	30	88	14	84	94	0
Metatarsals	7	0	21	0	34	100	0	29	88
Mandibles	4	0	50	38	100	50	32	82	0
Antlers	7	0	29	25	43	88	53	12	0
Ribs	13	0	100	94	63	100	100	100	0
Percoirs	1	0	0	0	13	16	0	18	88
Endscrapers	2	0	0	25	46	13	42	53	50
Burins	7	0	0	33	34	30	21	76	100
Cores	7	13	0	15	70	45	0	53	0
Bladelets	100	100	0	100	0	0	0	0	25

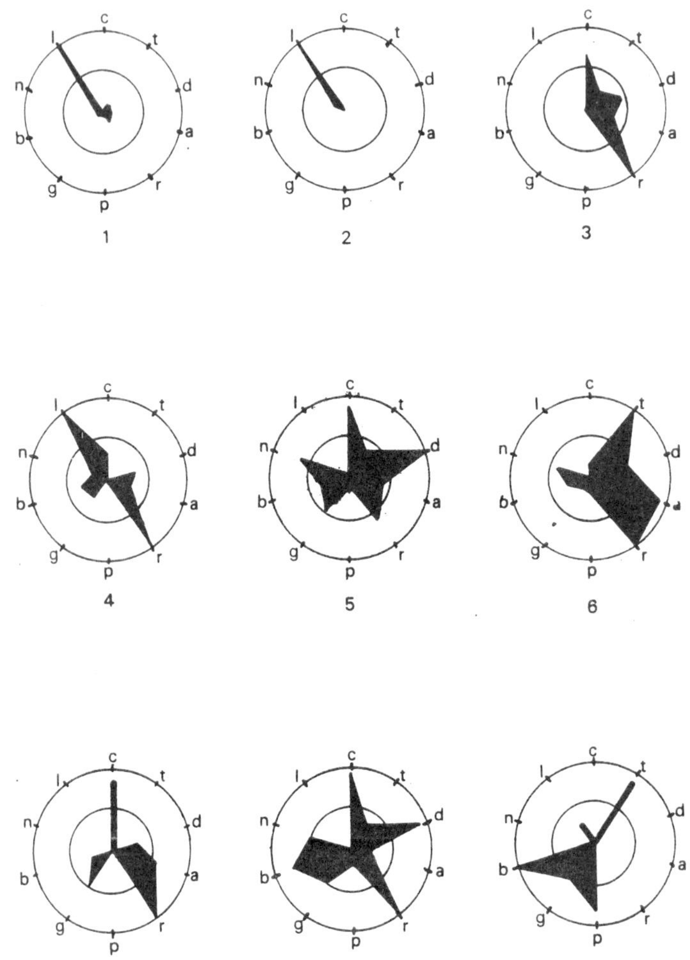

Figure 37. Circular Graphs for Zone Maximum Standardization of Zone Contents when Nine Zones Are Considered for Pincevent Section 36. Radii correspond to artifact classes as follows: (c) metacarpals, (t) metatarsals, (d) mandibles, (a) antler, (r) ribs, (p) percoirs, (g) endscrapers, (b) burins, (n) cores, (l) bladelets. Circumference of circle represents class maximum value of 100.

It should be pointed out that no lithic class has its maximum frequency in an isolated unit. Bladelets are found in greatest numbers in zones 1 and 4, while cores are most abundant in units 5 and 6. When the total number of bladelets occurring in and around hearth V105 are added together, including those from two isolated clusters and those from two zones, over 89% of the bladelets mapped at Pincevent can be associated with this one feature. This impressive dominance of the V105 assemblage by a single class may account for some of the differences in hearth zone contents among the three features.

Comparing results from the two transformations yields similar observations to those set forth for the six zone solution. Bladelets and ribs, when present, dominate a cluster zone's content since they tend to occur in large numbers. Similarly, these artifact classes also contribute to the overall relative abundance of a cluster zone. In some places, ribs and bladelets define a zone profile, yet they have different distributions. Bladelets seem to be deposited into small pockets, and these pockets contain large quantities of the class. Concentrations of bladelets occur in hearth-centered zone 1, but they are usually found away from hearths, either in rich zones near but not on the features or in isolated clusters. Ribs occur in hearth zones or in zones near hearths. They never form isolated clusters, and occur in dispersed masses rather than small pockets. Other artifact categories are found mixed together in rich cluster zones.

Small zones seem to be divided into two types: those dominated by bone materials and those with a lithic content greater than bone. Isolated clusters are exclusively lithic in content, with a high percentage of all bladelets occurring in isolates removed from the hearth areas. Cores are also found in isolation, particularly in the southwestern part of the site where they seem to associate with a small feature.

The general distinction made at the six zone level between dense and material-poor zones is maintained when nine zones are considered. Dense zones are found in close proximity to the three principal hearth features. Sparse zones and isolated clusters occur away from these areas, and frequently center on smaller features--cup hearths, stains, etc. The three richest zones (4, 5 and 6) are found in the northern part of the site centered on the two hearths and in a zone between them. These have similar profiles, although the zone associated with V105 is very rich in bladelets and poor in metatarsals. Zone 8 represents a less dense version of these zones and is centered on the southernmost hearth.

The analyses carried out here provide a basis for defining certain relationships which seem to represent an underlying structure to the deposits uncovered at Pincevent. These relate location and context to material abundance and to the content of specific areas within the site. In the following discussion, the results of rank matrix analysis will be combined with content studies based on maximum value transformations. Two aspects of spatial variation at Pincevent will be considered in this synthesis. First, the nature of spatial structure in deposition will be examined. Then, the problem of artifact association will be discussed in light of disposal structuring. In the final section, the deposits uncovered at Pincevent will be interpreted in terms of both the original conclusions proposed by Leroi-Gourhan and Brézillon and the patterning defined here.

DISCUSSION AND CONCLUSIONS

In certain areas at Pincevent, e.g., around the northern hearth features, materials were deposited in massive piles that seem to contain all artifact classes in quantities which reflect their overall abundance at the site. That this condition exists is evidenced by the intercorrelation of artifact classes at each clustering level produced by the k-means pattern search. Certain artifact classes do not seem to be part of this pattern, however. The examination of cluster zone composition based on maximum value profiles reveals that backed bladelets not only are distributed in a different spatial pattern than other classes but also have a distinct structure to their deposition.

The plotting of ranks at their k-means centroid locations points to a relationship among individual artifact classes of similar relative abundance. Spearman's rank order correlation statistics show that nearly all material categories are intercorrelated, despite progressive refinement of the spatial units being considered. Still, a difference between small cluster zones defined at high scales of refinement, i.e., at the nine zone solution, and the larger, hearth-centered masses seems to be present. Zone and class maximum proportions allow profiles of zone contents to be produced, displaying the composition of individual zones in easily compared graphs. These show that dense zones are essentially similar in content. Smaller zones are less consistent in terms of composition, and show a trend towards separation of bone and stone materials. However, no clear associations among specific artifact categories were found that might reflect activity patterning at Pincevent. Isolated clusters of bladelets, always found in large numbers, and cores do suggest spatial segregations of the activities involved in the production and/or use of these materials.

The zone maximum transformations for dense deposit areas are consistent not only within a clustering level but also between levels. Zones 2 and 3 at the three zone level, zones 3, 4, and 6 at the six zone level, and zones 5, 6, and 8 at the nine zone solution have consistent graphic profiles based on class maximum transformations. These zones represent materials deposited at hearth features T112 and L115, and in the area between the northern features. If, as Leroi-Gourhan and Brézillon suggest, the refuse piles focused on hearths were the result of a suite of "domestic" (within habitation structure) behaviours, then the correspondence of a "dump" area, i.e., that between the northern hearths, with the domestic refuse heaps is interesting. More importantly, assuming that hearth refuse results from in situ or nearby activities, the presence of a similar mass away from any hearth implies redeposition of domestic area refuse. A second such zone defined at the nine cluster level northwest of V105 (zone 1) supports the implication of refuse disposal behaviours conditioning spatial organization. Differences between all of these zones and those centered on V105 itself seem to relate to the large numbers of ribs and bladelets present in the northwestern hearth area.

Artifact association at Pincevent may be due to disposal processes structuring refuse deposition. It still is possible that in smaller cluster zones more restricted groups of objects can be related to activity sets. However, as was discussed above, no pattern or association is discerned in the peripheral zones. Moreover, sample sizes are very small in every case, precluding the use of statistical methods for pattern recognition. We are left with the possibility of accepting the outlying zones as they are, i.e., viewing each as reflecting "extensive use spaces" as defined by Binford, but this is a dangerous and unfounded assumption. If the presence of particular artifacts in a dense zone is predicated upon the overall abundance of material there, then this process may operate on smaller units as well. In other words, it is possible that variability in small units is stochastic and relates to sampling error.

We must interpret the Pincevent distributions in terms of disposal behaviours rather than in terms of activity specific areas. Important factors here are the social organization of the human group(s) occupying the site, as well as particular disposal behaviours that might explain the variation in spatial distributions noted between levels. Under this latter category, such factors as drop rates (Ammerman and Feldman 1974) might explain some of the variation in lithic objects found in peripheral zones, since the tool forms found there--scrapers and burins--are the most abundant excepting bladelets. A stochastic model for loss rates would predict higher numbers of these objects in low use areas. A second disposal mode involves the desire to keep domestic, or high use, areas clear of refuse so that other activities can be carried out there. Systematic cleaning and removal of garbage from hearth areas would result in dumps away from "intensive use spaces" that mirror the suite of behaviours occurring near the hearths. Two such areas seem to be present at Pincevent, and both of these are closest to the northern features.

Aspects of social organization at Pincevent suggested by the foregoing analyses conform to those presented by Leroi-Gourhan and Brézillon. The original interpretation of three habitations centered on major hearths is supported by the three zone solution. Fan-shaped refuse disposal away from the interior areas of each structure is also indicated, although the structure of those deposits can be elaborated. Certain amounts of material, indeed the richest masses, remain at the features. Disposal areas away from hearths in the northern habitations suggest either that these two occupations were of greater duration or that domestic activities were performed there more intensively than was the case for L115. That the activities performed within the habitation centered on V105 were somewhat different than in T112 and L115 is indicated by the concentration of bladelets in that area. In other respects, the three habitations are similar in artifact composition if not deposition intensity.

This pattern argues for the presence of three more or less equivalent productive units. Backed bladelets present something of a problem within this distribution, since they are over-represented in one unit. But given the tendency for this class to concentrate in small areas both within the V105 area and in smaller peripheral areas, they may indicate a particular production or use related to this one habitation unit. Even at the three cluster level, bladelets form an isolated cluster to the northwest of V105 containing 52.82% of all objects in the class. This concentration suggests a specific activity for this area, e.g., the production of bladelets.

Rib fragments also concentrate in the V105 area, with over 63% located there. However, this may be related to the intensity of pedestrian traffic around the V105 hearth, since similar numbers of reindeer (c. 20) are postulated for both northern habitations as MNI estimates (Leroi-Gourhan and Brézillon 1972: 161). Another mechanism that could account for the difference in rib numbers between habitation units is social in nature. Rib meat is often the most desirable of all butchered parts when fresh (see Binford 1979), and variation in meat distributed to various social units might account for the dominance of ribs in V105. The latter explanation implies the simultaneous occupation of the two northern units, but this has already been suggested by Leroi-Gourhan and Brézillon based on refitting of cobble fragments and characteristics of the distributions themselves.

The possibility of differential meat distribution is made more plausible when the nature of site use is considered. Pincevent is most likely a hunting locus, given the site's proximity to the Seine and the apparent short-term occupations represented. Leroi-Gourhan and Brézillon suggest a duration of at most five months, based on the sequencing of tooth eruption in reindeer mandibles. But the number of refuse piles identified here--only two away from the hearths--suggests limited maintenance of habitation areas if the amount of garbage still present at the hearths is an indication of "acceptable" preabandonment levels. These may have been greater than day-to-day accumulations but surely not overwhelmingly so. The indication from spatial analysis is of a short-term hunting camp within which cooperation between at least two, and possibly three, groups occurred; food distribution was probably a coincident element in the cohabitation context.

Some other evidence that supports such a contention is a separation of animal bones by age between the two northern structures. V105 contains a number of animals under six months and none between six and twelve months old. T112 has no bones that can be assigned to an animal less than a year old. If these units were occupied simultaneously, then differential resource use is indicated based on animal age, and a differential distribution of body parts in more mature animals--resulting in a preponderance of ribs in one unit--may be a part of the food-sharing process. It is possible that the differences in age distribution between these habitation units is linked to the predominance of bladelets in one of them, e.g., the processing of young animal skins into clothing. However, the site was probably occupied during the summer months, when caribou skins are least desirable (Skoog 1968).

To summarize the interpretations reached here based on the spatial analysis of the Pincevent materials, the original conclusions of the excavators are supported. There appear to have been three occupation units, at least two of which (V105 and T112) were there simultaneously. The habitation areas seem to have been occupied by more or less equivalent production units who carried out most of their activities in the shelter of warming fires. Part of their behavioural repertoire included the systematic cleaning of habitation areas, and this resulted in the creation of dump areas which show similar composition to those at the hearths themselves. A few activities, involving the manufacture and/or use of backed bladelets and probably flint knapping, were carried out at some distance from the central habitations. Most other activities occurred within the occupation areas. There is some evidence for food sharing with differential access to certain body parts, including a high incidence of a desirable part, the rib slab, in one habitation. A relatively short-term occupation is suggested by the presence of few refuse piles resulting from disposal behaviours and rich "preabandonment" masses left at the hearth loci. This also may reflect intense activity at the hearths prior to abandonment, since raw animal refuse becomes unbearable after even a few days and the hearthside accumulations are sizable.

The foregoing conclusions are tentative at best, merely providing a "good fit" to the patterns noted analytically. Still, the patterns themselves have important implications for further spatial analysis. The presence of refuse disposal behaviours precludes the search for "tool kits" in the Pincevent spatial distributions. Given that this site is one of the most remarkably preserved records of Paleolithic habitation known, the prognosis for finding tool kits in disturbed or more complex depositional environments is grim. However, one question remains: are the behaviours inferred based on the Pincevent analyses always operative, or are there conditions under which activity specific areas maintain their spatial integrity? Possible conditions might be biocultural, i.e., disposal behaviours may not have been part of the human behavioural set throughout its history. The site context may influence the structuring of deposition in behavioural terms. It may be that certain activities carried out at Pincevent required the removal of refuse materials from activity areas. The nature of the site location itself may have conditioned deposition behaviour, and other location types may have other constraints. With these problems in mind, Paleolithic habitation floors uncovered at the rockshelter site Le Flageolet I were studied. These analyses are presented in the sections which follow.

CHAPTER 5

INTRODUCTION TO LE FLAGEOLET I

SITE LOCATION

Le Flageolet is a rockshelter site located near the small village of Bézenac in the Dordogne Department of southwest France (Figure 38). One of many known cave and rockshelter sites in this part of the Dordogne River Valley, Le Flageolet is an erosional shelter within a small outcrop of the Upper Cretaceous limestones that form the underlying lithostrata for much of the Perigord region (Rigaud 1982: 286). These limestones, particularly in the area of St. Cyprien near Le Flageolet, have undergone significant alteration by tectonic forces in the past (Fenelon 1951). A transverse fault bisects the cliffs in the center of Le Flageolet.

Two principal areas of rock erosion form two rockshelters at Le Flageolet, and these have their interface at the fault (Laville 1975: 283). One shelter is called Le Flageolet I and has a western orientation. Sediments within Le Flageolet I date to the Wurm III stadial in the Alpine scheme. The second shelter is referred to as Le Flageolet II, faces to the south-southwest, and contains sediments dating to the Wurm IV. The two rockshelters meet at a sharp angle in the center of the outcrop.

Le Flageolet lies at the lowest point within the Upper Cretaceous limestone outcroppings in this part of the Dordogne Valley (Rigaud, *ibid*: 288). To the north of the site, a small valley (also formed, in part, by the transverse fault) climbs gently to the surrounding plateau. In this part of the Dordogne (called the *Sarladais*), upland areas are blanketed by a mantle of Eocene siderolites originating outside the region. These sediments were probably deposited on the plateau at the Plio-Pleistocene boundary and have mixed with local decomposing bedrock in some places (Texier 1979). The mantle is referred to as the "Siderolithic" and often contains chert nodules eroded out of the underlying Cretaceous limestone. Thus, chert was available locally throughout the later Pleistocene both within some limestone outcrops and scattered over the landscape in the plateau sediments.

The fault passing through Le Flageolet continues across the river valley to the south. Where it crosses the river, different elevations on either side of the fault have caused coarse gravels to accumulate. These

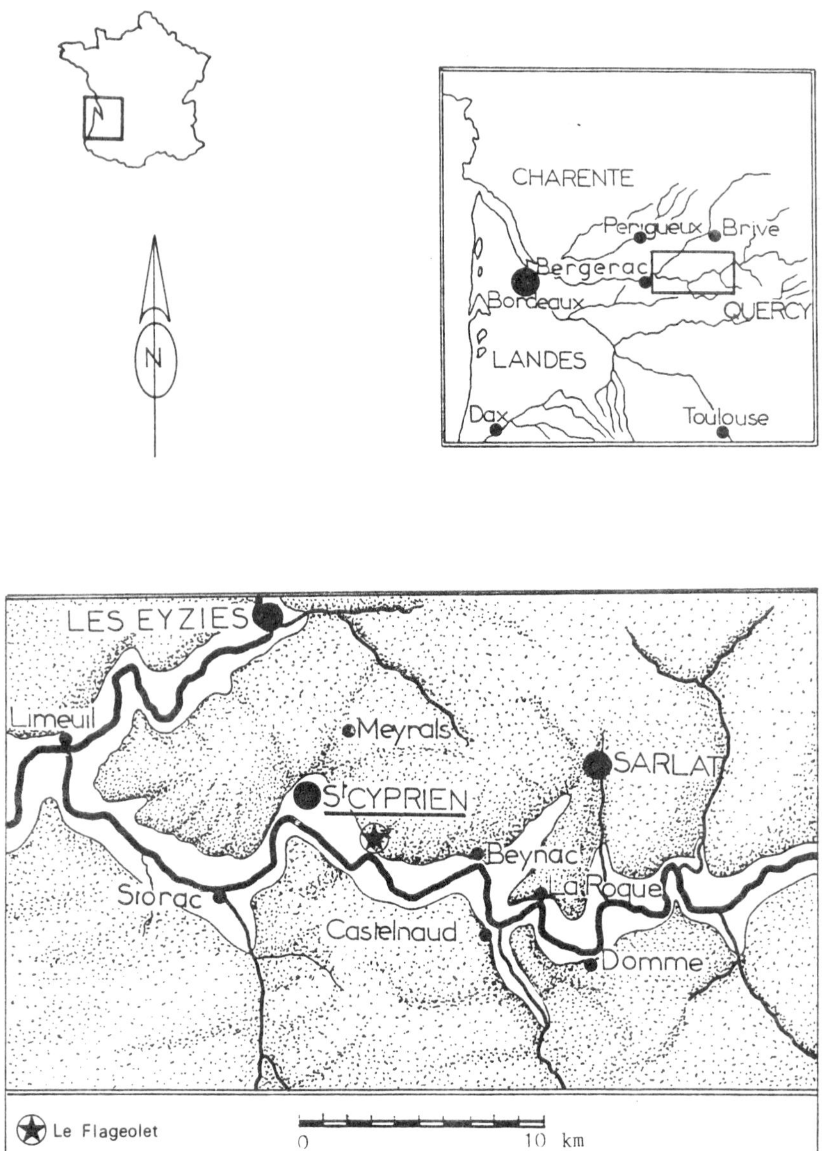

Figure 38. Map of Southwestern France Showing Location of Le Flageolet I and Surrounding Region. Star marks the site location. Reproduced from Rigaud 1982. (Drawn by J-Ph. Rigaud).

gravels create a shallow bar across the river--a ford. Since much of the load carried by the Dordogne derives from the igneous Massif Central to the east, the gravels at the ford include quantities of quartzite and basalt. As will be seen, these materials were exploited by the inhabitants of Le Flageolet.

That the fault influences ford formation and has some antiquity makes it another kind of potential resource for paleolithic hunters. Fords like this one would have provided river crossings for migrating caribou herds during the Pleistocene, assuming that the animals migrated at that time and that their direction of movement was east/west. White (1980) has argued that paleolithic site location was, in part, determined by the locations of fords and their potential as such resource spaces. While there is, at present, no evidence confirming the necessary animal behaviour, Le Flageolet would have served admirably as a watch and access station for the exploitation of caribou crossing at the ford below. In addition, the site location provides easy access to the plateau via the gently sloping valley above the cliff. The cliff itself is easily traversed today near Le Flageolet, but there is no reason to assume that this was the case 25,000 years ago.

Still, in all probability, Le Flageolet provided access to a variety of ecozones (plateau, interflueve, valley slopes, river bottom), and lithic raw materials were locally available in the form of chert nodules. These benefits seemed to have attracted human groups to the rockshelters throughout the later Pleistocene. In the two sectors of the site, an almost complete sequence of Upper Paleolithic archaeological traditions are present in the shelter sediments (Rigaud 1969, 1982; Laville, Rigaud and Sackett 1980).

HISTORY OF WORK ON THE SITE

After discovering Le Flageolet in 1965, Jean-Philippe Rigaud, Regional Director of Prehistoric Antiquities for the Aquitaine, carried out extensive excavations in both areas of the shelter. Rigaud's work began in 1966 and continued every summer without interuption until 1984. A series of articles have been published concerning discoveries at the site (Rigaud 1969, 1970, 1976a, 1976b, 1978a, 1978b, 1980, 1982). A team of specialists has undertaken intensive analysis of the local environment during the paleolithic occupations of Le Flageolet. These studies include faunal analysis (Delpech 1970, 1975; Delpech and Rigaud 1974) and sedimentology (Laville 1970, 1975). Although the majority of these publications are in French, the site has received some treatment in English, including a general description (Laville, Rigaud and Sackett 1980), an analysis of assemblage variation

(Laville and Rigaud 1974), and spatial studies of several occupation layers (Rigaud 1978b; Simek 1981).

PLEISTOCENE SEDIMENTATION AT LE FLAGEOLET

Rigaud divides Le Flageolet into two "sites," corresponding to the two shelters separated by the fault (1982: 290). South of the fault, a sequence of at least nine occupations are discerned within the Würm III sedimentary deposits. The northern part of the shelter contains a sequence of at least three Magdalenian occupations within the Würm IV sedimentary series (Rigaud 1969, 1970). Athough the shelter is continuous, Rigaud distinguishes two sites (Le Flageolet I and Le Flageolet II) based on geologic processes that resulted in sediments deposited during different stadial periods in the two sectors.

During the Würm III, it seems likely that the entire shelter at Le Flageolet was filled with sediments. These derived both from the siderolithic mantle on the plateau above and from cryoclastic weathering and subsequent chemical decomposition of the limestone outcrop itself. Filling of the shelter was gradual during the Würm III, and human groups occupied the shelter at various times during the stadial. These groups made two different kinds of stone tool assemblage; earlier peoples produced stone tools typical of the Aurignacian tradition, and later groups produced an Upper Perigordian industry. At the end of the Würm III, a substantial increase in local and regional temperature and humidity accompanied the advent of the Würm III/IV interstadial. Water from the plateau coursed through the valley above Le Flageolet and cascaded into the main river valley at the site's central fault. The resulting waterfall eroded away the sediments north of the fault. This erosion emptied the sector designated Le Flageolet II of Würm III sediments.

The onset of the Würm IV stadial witnessed a return to severe cold and dry climatic conditions, and sedimentation resumed in the northern sector of Le Flageolet. Magdalenian groups occupying the site were confined to Le Flageolet II by the surviving mound of sediments south of the fault. They nevertheless built up a substantial midden deposit in Le Flageolet II, and today, sedimentary profiles nearly ten thousand years apart in age have similar absolute elevations over only eight meters of distance. Because the spatial studies presented here are concerned only with occupations in Le Flageolet I (Aurignacian), detailed discussion of site characteristics will be limited to deposits from the southern part of the rockshelter.

PALEOENVIRONMENTAL RECONSTRUCTION

Sedimentology of Le Flageolet I

The sediments that fill Le Flageolet I span the Würm III stadial period and provide data for reconstructing certain environmental parameters that might have affected human life during the Upper Pleistocene. This sedimentary record is being analyzed by Henri Laville of the Institute of Quaternary Studies at the University of Bordeaux. Preliminary findings are discussed in Laville's thesis (1975) and in articles by Rigaud (1969, 1980, 1982). The summary of geologic information presented here relies entirely on these sources as well as on personal discussions with both scholars.

The rockshelter at Le Flageolet was artifically walled in for use as a house during the medieval period. This situation is not at all rare in the Perigord, which is famous for the use of shelters and caves in dramatic architectural constructions that serve as houses, barns, animal pens, etc. During the middle ages, and perhaps under pressure from raiding Viking maurauders (who ascended the navigable waterways of southern France in their quest for booty), the farmers and herders of the rural Dordogne Valley sought refuge in the defensible cliff faces (Scargill 1974). The actual date of wall construction at Le Flageolet is unknown, but it was likely to be rather late compared to to other "troglodite" settlements in the region (e.g., La Rocque St. Christophe). An official communal road passes just at the foot of the shelter, and this was used as a principal route between the towns of St. Cyprien and Beynac. As early as the 18th century, Le Flageolet was used both as a habitation and as an inn along the road (Rigaud, personal communication). When construction of the house occurred, it resulted in the removal of as much as 2 m of sedimentary deposits in Le Flageolet I during wall raising and floor levelling operations. Below the house floor, which was of swept earth, a long sequence of Würm III sediments survived, protected by the residents above.

Beginning with the uppermost and most recent surviving sediments, Laville (1975: 286-9), working in collaboration with Rigaud, distinguishes the following sequence of geologic beds or Couches:

> Couche I: Rounded limestone gravels in a sandy-silt matrix. Within this loose and friable matrix occur some angular limestone rocks resulting from cryoclastic weathering (fr. *plaquettes*).
>
> Couche II: This is a discontinuous Couche made up of sandy-clay lenses without the gravels found in Couche I.

Couche III: This bed is similar in composition to Couche I in areas where both are found in the same profile. Gravels dominate the sediments, again within a matrix of sandy-silts. At the base of Couche III, there is an increase in the proportion of plaquettes.

Laville discusses these three strata as separate natural beds, but their discontinuity over the site and the lenticular nature of Couche II within Couches I and III have prompted Rigaud to consider them as a single geologic unit (1976c). These three upper Couches are illustrated at the top of Figure 39, a sagittal profile observed in the center of Le Flageolet I. The location of this and other profiles illustrated here (Figures 40 and 41) are indicated on the site plan presented as Figure 42.

Below Couches I-III, Laville defines a series of eight more natural strata:

Couche IV: (20 cm maximum thickness). This is a rather homogeneous layer of sandy silt containing rare angular limestone blocks of medium size (fr. eboulis). The boundaries of this layer are sharp throughout the site as is apparant on the profiles (Figures 39, 40, and 41).

Couche V: (30 cm maximum thickness). Couche V represents an abrupt change from both higher and lower sedimentary units. The layer is composed of small plaquettes lying, for the most part, horizontally. Particularly in the south of Le Flageolet I, the plaquettes are so dense that there is virtually no matrix within Couche V.

Couche VI: (25 cm maximum thickness). This Couche is composed primarily of silty sands mixed with occasional larger eboulis.

Couche VII: (40 cm maximum thickness). Couche VII is charaterized by small plaquettes in a sandy-silt matrix. At the top of the Couche, these are mixed with rounded limestone gravels (Rigaud 1980). Laville considers the plaquette component of Couche VII more developed towards the back of the shelter (i.e., near the wall) and distinguishes three subunits within the layer. The upper part of Couche VII consists of "an alignment of angular plaquettes rarely over 30 mm in diameter" (1975: 287). The middle part of the Couche also contains plaquettes, but these are larger than in the upper subunit. The base of Couche VII contains a significant eboulis component with some rocks attaining considerable size.

Couche VIII: (100 cm maximum thickness). This Couche, essentially continuous with Couche IX below, comprises angular limestone eboulis in a matrix of fine sandy clay-silts. There is some local variation in composition within the shelter, both horizontally and vertically. Nevertheless, Couche VIII is truely a rock layer. In some areas, thin sand lenses occur within Couche VIII, but these are discontinuous and cannot be related from one area of the site to another.

Couche IX: (100 cm maximum thickness). At the beginning of excavations, Rigaud made a stratigraphic distinction between Couches VIII and IX. Laville's sedimentological work and more extensive areal excavations have shown that there are no real, consistent differences in sedimentary composition between the two layers (Rigaud 1980, 1982). Laville discusses Couches VIII and IX as a single level in his thesis (1975), and the description of Couche VIII presented above suffices here. In some areas of the site, Couche IX contains larger, more block-like eboulis that does VIII. This is particularly the case at the base of the stratum.

Couche X: This layer is composed of huge limestone blocks, often many cubic meters in volume, that are scattered over the shelter floor. Laville defines Couche X as a separate geologic layer because it seems to represent a single major roof collapse early in the Wurm III. There are indications that subsequent cryoclastic weathering resulted in further rock fall, but these later events were in no way as intense as the massive collapse resulting in Couche X. The blocks making up Couche X had important effects on later sedimentation in Le Flageolet I. In Couches IX and VIII (and to some extent VII), sedimentation occurred within and around the blocks. Thus, these layers are confined to narrow cracks. This fact, coupled with the size of coarse elements in these strata, produced layers that are thick (see above) but that were not necessarily deposited over long periods of time. After the deposition of Couche VII, however, the cracks were mostly filled, and the internal topography of Le Flageolet I was more or less level. Couches I to VI were laid down as horizontal beds with a slight slope towards the fault (i.e., from south to north).

Couche XI: (20 cm maximum thickness). This is a thin layer resting on the shelter floor (bedrock). The matrix is yellowish clay-sands resulting from in situ chemical weathering of the limestone. The blocks of Couche X fell directly onto Couche XI in most parts of the shelter.

Climatic changes indicated by analysis of these sediments are discussed by Laville (1975; Laville, Rigaud and Sackett 1980). He sees the composition of Couches I-III as indicating a mild and humid climate that, "produced sporadic manifestations of solufluction or runoff" (Laville, Rigaud and Sackett 1980: 256). Couche IV shows heavy particle weathering but contains few limestone slabs. Indications here are for a depositional climate similar to that in Couches I-III--warm and humid with little cause for cryoclastic weathering. A radical change in local climate is evident between Couches IV and V. Some solufluction is indicated at the upper surface of Couche V, probably relating to the gradual warming trend that resulted in Couche IV deposition. For the most part, Couche V comprises sediments formed and deposited during very rigourous (cold and dry) climatic conditions. Couche VI, even though it does contain some eboulis, seems to represent another mild period during the Würm III as evidenced by a substantial fine fraction. There are also indications of chemical weathering of the sediments--more evidence for mild conditions. Another cold and dry period is indicated by Couche VII deposits. However, there is evidence here for unstable, i.e., fluctuating, climatic conditions. While the overall picture is of a cold and dry phase, lenses of gravel suggest brief warmer periods (Laville 1975: 291).

Unfortunately, detailed sedimentological analysis of Couches VIII and IX have not been completed. The climatic interpretations presented below are, thus, preliminary and paraphrase those made by Rigaud (1980, 1982: 295). Still, some tentative correlations between the Le Flageolet I sequence and other Perigord sites can be made based on Laville's "chronostratigraphic" method. This technique is somewhat analogous to dendrochronology but uses the sequence of climatic changes revealed by sediments (Laville 1975; Laville, Rigaud and Sackett 1980). Using this method, some indication of the climatic regimes extant during deposition of Couches VIII and IX can be extrapolated.

Rigaud (1980) relates the sediments of Couche V to a well known Würm III marker bed first identified in Couche 9 at the Abri Facteur (a. k. a. La Foret) during Delporte's excavations (Delporte 1968). This bed reflects a period of brief yet very severe climatic conditions towards the end of the Wurm III stadial. Among archaeologists, this cold, dry period is known as the coupe de froid (cold snap) or as the "Tursac oscillation". The correlation of Couche V at Le Flageolet I with Couche

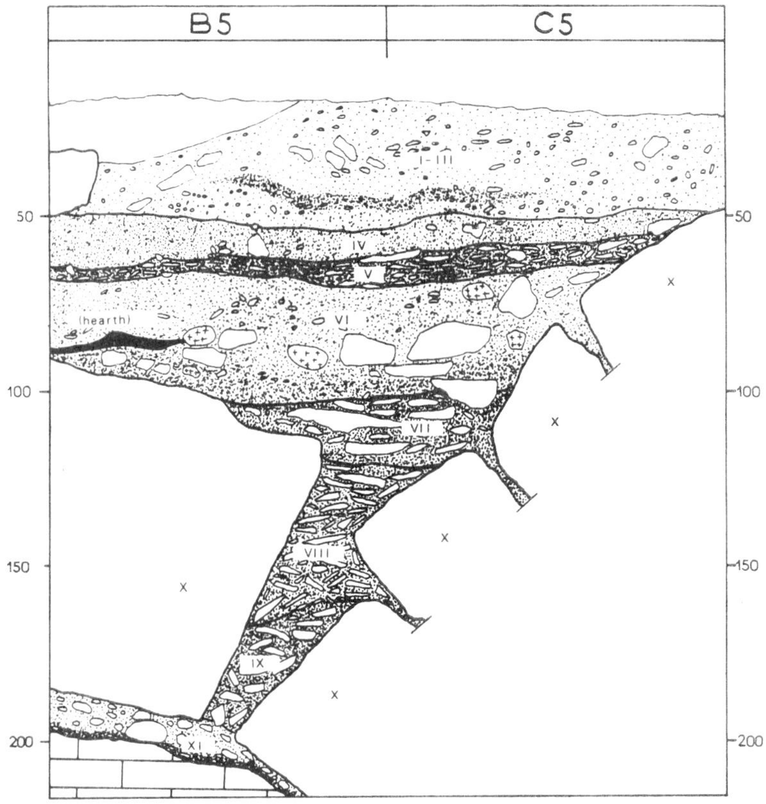

Figure 39. Sagittal Profile of Le Flageolet I Sediments Taken at Grid Units B5/C5 Redrawn from Laville and Rigaud's field observations.

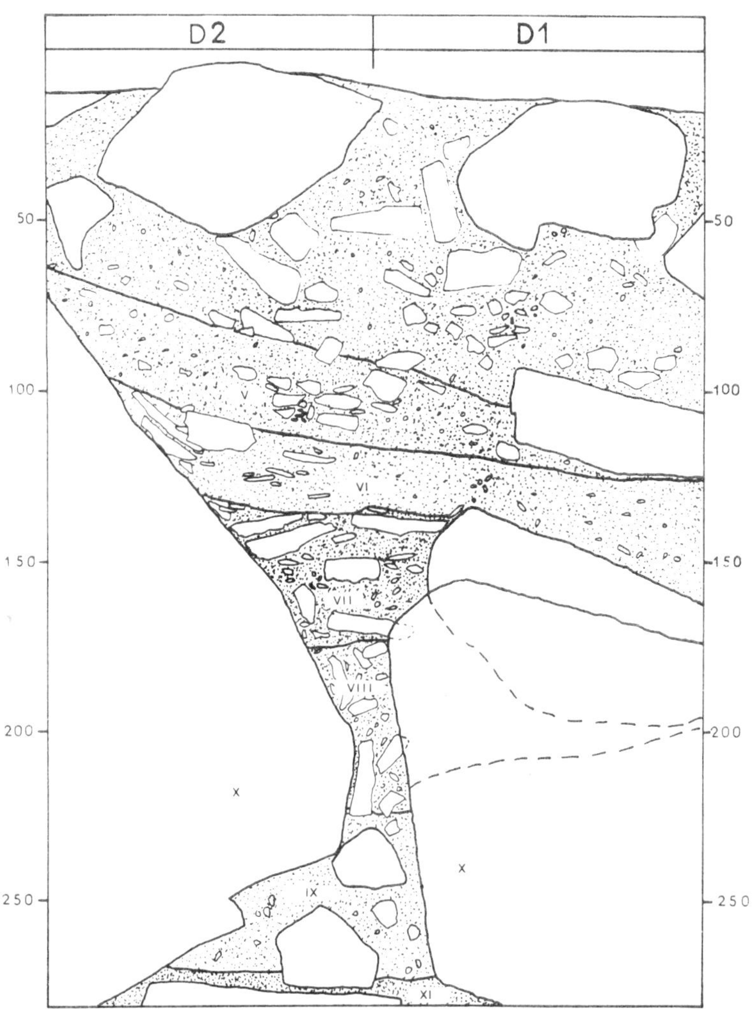

Figure 40. Frontal Profile of Le Flageolet I Sediments Taken at Grid Units D2/Dl Redrawn from Laville and Rigaud's field observations.

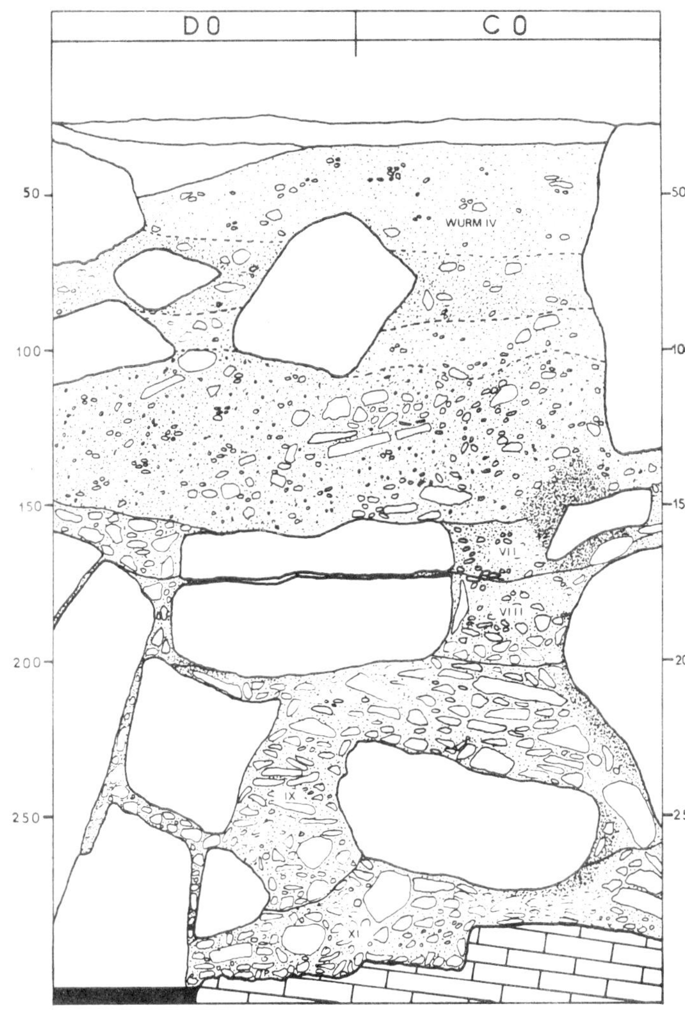

Figure 41. Sagittal Profile of Le Flageolet I Sediments Taken at Grid Units C0/D0 Redrawn from Laville and Rigaud's field observations.

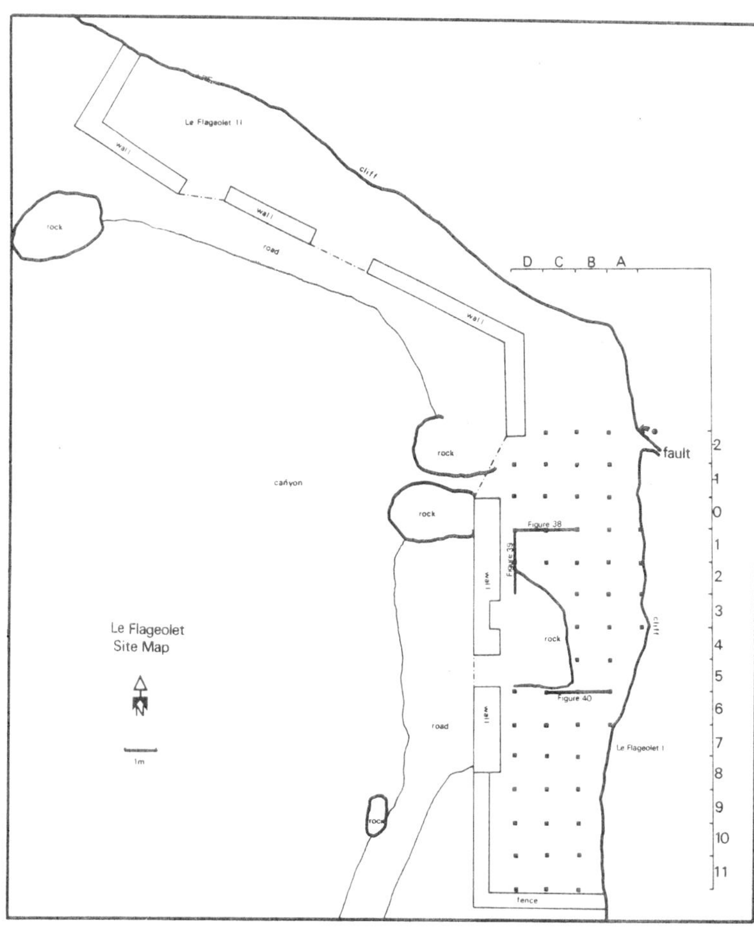

Figure 42. General Plan Map of Le Flageolet. The grid system employed in excavations is illustrated, the locations of profiles shown in Figures 39-41 are indicated, and the medieval walls are outlined.

9 at the Abri Facteur anchors the former sequence into Laville's regional chronostratigraphy. Based on this, the relative position of the lowest layers can be approximated.

Rigaud (1982) suggests that Couche VIII may correlate with Couche 4 at the Roc de Combe (Bordes and Labrot 1967). This would represent a mild and humid climatic regime for deposition. But the eboulis content of Couche VIII suggests some rather extreme (perhaps seasonal) fluctuations from cold to warm humid conditions. It is this kind of fluctuation that usually produces maximum cryoclastic weathering. Since no discernable difference in sediments exists between Couches VIII and IX, we can only guess the precise nature of the climate prevalent during Couche IX deposition. Both couches may have been formed under mild and humid regimes characterized by marked seasonal fluctuations in temperature. Together, Couches VIII and IX compose a deep deposit, usually implying long periods of time. Laville's chronostratigraphic correlations reveal no very long period of mild, humid climate during the Wurm III. However, remembering that Couches VIII and XI are located within narrow cracks between blocks and are made up of large particles for the most part, the volume of sediments composing the layers is not all that great. They might well have been laid down in a short time, and Rigaud's correlation with Roc de Combe may not be far from wrong.

Still, as Rigaud himself states, only further and more detailed sediment study will resolve these issues. As will be discussed below, the archaeological materials contained in the lower strata complicate rather than alleviate chronological problems. However, preliminary faunal analyses do contribute information that aids in environmental reconstruction in the lowest Le Flageolet I layers.

Faunal Analysis of the Le Flageolet I Aurignacian Couches

The fauna found in association with archaeological materials at Le Flageolet I was (and is still being) analyzed by Françoise Delpech of the Institute for Quaternary Research at the University of Bordeaux (Delpech 1975). Since the occupations of concern here are limited to the Aurignacian surfaces contained in Couche VIII, only faunal data from layers VIII and IX will be discussed in detail. Delpech analyzed the two Couches as a single unit since they were by then considered a unitary "sedimentary ensemble" (*ibid*: 62). In any case, the fauna seems to be homogeneous throughout the layers. They also support the tentative climatic interpretations offered by Rigaud based on chronostratigraphic correlation.

Delpech emphasizes the "temperate" nature of the megafauna recovered from Couches VIII and IX (Table 17). Red deer, roe deer, boar, auroch, horse, and donkey are all present in the assemblage. Most of these creatures are associated with various biomes exant under mild and humid climatic regimes. Interestingly, species normally associated with more rigourous conditions are also present (caribou, ibex, and chamois), but these animals may have remained in the region during warmer phases of the Wurm in restricted micorenvironments. Ranges of microclimatic variation are great in the Perigord region even today, and there is no reason to expect less variability in local conditions during the Pleistocene (cf. White 1980: 71).

Based on his work at the Abri Pataud, Wilson (1975) argues that regional conditions in the Dordogne during the Wurm were similar to those present today at an altitude of ca 1000 m in the Massif Central. That environment is characterized as a grassland biome with limited stands of arboreal plants in low-lying, protected areas. Wilson suggests that similar wooded refugia would have existed near south-facing cliff exposures along the Dordogne Valley during the Wurm, since temperatures there are consistently higher than in surrounding, more-open areas (ibid: 61). Le Flageolet represents such a south-facing cliff. Given the wide range of microenvironments accessible from Le Flageolet, including the upland plateau, the valley bottom, and the cliff walls themselves, a wide range of animal species might be expected in any faunal assemblage produced by hunters using the site. As Delpech's analyses show, this is, indeed, the case; species preferring both mild and wooded ecosystems and more open and rigourous conditions are represented in the Couche VIII/IX fauna.

Microfaunal remains (mostly of rodents) are also abundant in the VIII/IX sediments (Table 18) None of these small bones have been obviously burned, so their association with the human occupations of the shelter cannot be firmly established as yet. Still, rodents are very sensitive to local climatic conditions, and their presence in these sediments can provide information concerning changing environmental conditions during the formation of the geologic layers.

Several species of fieldmouse (fr. campagnol) are present in Couches VIII and IX. The numerical dominance of the Grand Campagnol suggests that relatively warm and humid conditions prevailed throughout deposition of the layers (Delpech 1975: 66). A whole group of rodent species accompanies these fieldmice, including some that favor colder conditions (e.g., *Microtus nivalis*). Yet, warm loving species are clearly the most frequent. Small mammal remains in Couches VIII and IX indicate a generally warm and humid climatic regime during the formation of these levels.

TABLE 17.

Number of identifiable specimens for megafauna bone fragments recovered from Le Flageolet I levels VIII-1, VIII-2 and IX combined (after Delpech 1975: Table 10).

SPECIES	NISP
Cervus elaphus L.	10
Rangifer tarandus L.	8
Equus caballus L.	5
Bos sp.	2
Vulpes vulpes L. or *Alopex lagopus* L.	2
Capreolus capreolus L.	1
Rupicapra rupicapra L.	1
Sus scrofa L.	1
Equus hydruntinus Regalia	1
Mustela erminea L.	1
Mustela nivalis L. or *Mustela minuta* Pomel	1

TABLE 18.

Number of identifiable specimens for microfauna bone fragments recovered from Le Flageolet I levels VIII-1, VIII-2 and IX combined (after Delpech 1975: Table 11).

SPECIES	NISP
Indeterminate larger fieldmice	574
Arvicola sp.	127
Microtus arvalis Pallas or *Microtus agrestis* L.	49
Talpa europaea L.	11
Eliomys quercinus L.	8
Microtus nivalis Martins	7
Erinaceus europaeus L.	7
Citellus sp.	2
Apodemus sylvaticus L.	2
Microtus gregalis Pallas	1

Twenty-eight species of birds are part of the Couche VIII/IX faunal assemblage (Table 19). Delpech makes no ecological inferences based on these remains. However, the presence of at least three carnivorous species is noted (falcons). Frequencies of these rapacious predators are relatively high in both Couches VII and VIII/IX, and these sedimentary units also have the highest numbers of small mammal remains. It may be that both non-predatory birds and rodents were brought into Le Flageolet by birds of prey rather than as results of human behaviour. No choice between these two agents, hawks and humans, can be made at this time. It is interesting to note, however, that a negative correlation between micromammals and human occupation was observed at the cave site Abri Vaufrey (Rigaud, personal communication). This negative relation may reflect the avoidance of suitable habitations by birds of prey when humans are present. If this is so, the quantities of microfauna, and of avian predator remains, in Couches VIII and IX may indicate rather sporadic human presence during the formation of the layers. In any case, the bird species present in the assemblage are not inconsistent with the climatic reconstructions already offered. Again, a warm and humid regime is suggested with, perhaps, periodic fluctuations toward colder temperatures.

Finally, four fish species are represented at Le Flageolet I in Couches VIII and IX. These are trout, eel, perch, and pike; all are historically known from the Dordogne River below. A single species of salamander is identified, but the same animal crawls around in the shelter today, burrowing among the rocks. The antiquity of these amphibian remains is in doubt.

In summary, all faunal elements suggest a similar view of the region around Le Flageolet during the Aurignacian occupations of the site. A warm and humid environment is indicated, supporting a wide variety of animal species. Species favoring forested habitats are most common, with the red deer the dominant large mammal form. Species favoring cooler and/or more open habitats are also present with the forest forms, suggesting that the occupants of Le Flageolet I (at least during the deposition of Couches VIII and IX) used the site's central location to exploit a variety of microenvironments. In short, there is nothing in the faunal assemblage to suggest areally extensive or specialized hunting subsistence (e.g., the "reindeer economies" so often attributed to Upper Paleolithic peoples).

Palynological Analysis of Le Flageolet I Couche VIII

Studies of the pollen preserved in Le Flageolet sediments were carried out by Marie-Françoise Diot of the National Center for Prehistory in Perigieux. Her

TABLE 19.

Number of identifiable specimens for avifauna bone fragments recovered from Le Flageolet I levels VIII-1, VIII-2 and IX combined (after Delpech 1975: Table 12).

SPECIES	NISP
Perdix perdix L.	42
Coracia graculus L.	26
Coloeus monedula L.	21
Indeterminate smaller corvids	10
Falco tinnunculus L.	10
Aytha fuligula L.	7
Coturnix coturnix L.	7
Alauda arvensis L.	4
Tringa nebularia Grunn.	4
Indeterminate ducks	3
Crex crex L.	3
Indeterminate turdids	2
Turdus musicus L.	2
Anas sp. (teal)	2
Columba livia Gmel. or *Columba oenas* L.	2
Corvus corax L.	1
Indeterminate passerines	1
Turdus merula L.	1
Turdus viscivorus L.	1
Turdus pilaris L.	1
Indeterminate falcons	1
Falco columbarius L.	1
Anas platyrhynchus L.	1
Botaurus stellaris L.	1
Gallinago gallinago L.	1
Tringa erythropus Pallas	1
Vanellus vanellus L.	1
Hirundo sp.	1

examination of 19 samples from Couche VIII produced too few pollen grains to offer environmental reconstructions. Diot states,

> A pollen profile based on 112 pollen grains from the same altitude in Couche VIII, but from three different grid squares, yields results that are difficult to interpret [Diot 1981: 4]

Although herbaceous plants and ferns dominate the limited pollen sample, such a spectrum mirrors present conditions at the shelter. Given the poor pollen preservation in the sediments overall, and the possibility of contamination either prior to or after sample collection, Diot does not feel it prudent to offer reconstructions of Le Flageolet's local vegetation in prehistoric times.

Summary of Environmental Data From Couches VIII and IX

Based on analysis of sediments, large mammals, and small mammals, a generally warm and humid climate prevailed at Le Flageolet as geologic Couches VIII and IX were deposited. While all of these studies are incomplete, preliminary results agree with Rigaud's tentative chronostratigraphic interpretation of the layers. Variation in this general climatic situation is evident, both in the massive amount of cryoclastic debris incorporated in the sediments and in the presence of certain cold loving animals. These divergences from the general environmental picture might be explained as reflecting (1) microenvironmental variation over short distances in this ecologically diverse region, and (2) seasonal or more long-term fluctuations in climatic conditions that made the locale temporarily attractive to cold loving species and resulted in intensive mechanical weathering of the cliff. It should, of course, be noted that the ecological tolerances of Pleistocene animal species need not coincide with their present analogues. Still, the exploitation of various microenvironments by occupants of Le Flageolet I is suggested by the faunal diversity. Site location, providing access to both valley and uplands, favors such a locally intense and generalized pattern.

ARCHAEOLOGICAL STRATIGRAPHY

An extensive series of archaeological levels (fr. *niveaux*) was uncovered during Rigaud's fifteen excavation seasons at Le Flageolet I. Several discussions of the sequence have appeared in print (Rigaud 1969, 1976a, 1976b, 1980, 1982; Laville, Rigaud and Sackett 1980). The following outline of the archaeological stratigraphy depends heavily on the most recent synthesis, Rigaud's

1982 thesis.

The uppermost geologic layers contain at least five Upper Perigordian occupation surfaces. The latest occurs within Couche I-III. The assemblage from I-III contains 156 retouched stone tools. Of these over 30% are burins. A peculiar flat-faced burin (approximating the Raysse-Basseler type usually associated with the Perigordian V facies) is present in I-III, but this is different enough from the Raysse form to be classified as a more general "flat-faced burin". Gravette and Microgravette points, type-fossils for the Upper Perigordian (also called "Gravettian"), are also part of the assemblage. Rigaud assigns the Couche I-III assemblage to the Perigordian VI based on its composition and stratigraphic position.

Couche IV yielded 138 tools. Again, burins are the dominant form, comprising over 50% of the assemblage. The Raysse-Basseler burin is present in its typical form. There are no "Noailles" burins, another type-fossil for the Perigordian V, so Rigaud attributes the Couche IV assemblage to a generalized Upper Perigordian (1980: 11).

Couche V is a very rich surface currently undergoing extensive technological and spatial analyses. Over 500 tools were recovered from this thin occupation level, associated with a series of well-preserved features. Burins are, again, the most common assemblage component. Both Raysse-Basseler and Noailles burins are present, supporting Rigaud's suggestion that both were part of the same assemblage tradition. The industry from Couche V is attributed to the Perigordian V.

Couche VI contains at least one, and possibly two, archaeological levels; stratigraphic interpretation of this deposit has been difficult. A total of 531 tools compose the assemblage from Couche VI, and the proportion of endscrapers is close to that for burins. Both Noailles and Raysse-Basseler burins are present in the assemblage, along with the ubiquitous Gravette and Microgravette points. Two other tool types associated with Perigordian V facies are also present: truncated elements (blades with retouched truncations on both ends and usually steep lateral modification) and stemmed Font Robert points. These associations, also present in Couche VII below, are important since these tool types were thought to represent type-fossils for different chronological facies within the Perigordian V. That they occur within a single occupation at Le Flageolet calls their utility as chronological markers into serious question.

Couche VII is the richest and earliest Upper Perigordian level in the Le Flageolet sequence. The industry comprises 580 tools and a large amount of lithic waste. In contrast to the other Upper Perigordian levels at Le Flageolet, indeed to the Upper Perigordian generally, endscrapers are the dominant tool form in the assemblage. As was the case in Couche VI, the three Perigordian V chronological type-fossils (Noailles burins, truncated elements, and Font Robert points) are again associated in the same occupation.

Below the Upper Perigordian series, at least four Aurignacian levels have been excavated in whole or in part; two of these form the data base for the present analysis. For this reason, characteristics of the Aurignacian assemblages from Le Flageolet I will be discussed in more detail than was the case for the Perigordian industries. This discussion will provide a general background to the two assemblages to be analyzed in succeeding chapters.

As recently as 1980, Couche VIII was treated as a single occupation (Rigaud 1980), then containing a total of 244 tools. These materials, based on data recovered through the 1979 excavation season, included 81 endscrapers (31.3% of the assemblage). A Typical Aurignacian industry is present, as witnessed by the frequent Carinated and Shouldered scrapers (Figure 43 A and B). Another typical Aurignacian tool form, the "Busqued" burin is represented by a few examples (Figure 43 C) linking the site to other Aurignacian assemblages recovered from the Dordogne (but not the Vezere) Valley. Small, retouched microliths called Dufour bladelets are noted (Figure 43 D and E). Carinated burins, also an Aurignacian type-fossil, are present (Figure 43F), along with a series of more mundane burins (Figure 43 G and H). The assemblage from Couche VIII is completed by a series of truncated pieces (but not truncated elements) and by a variety of partially retouched and notched pieces.

As was discussed above, Rigaud suggests that the general characteristics and stratigraphic positon of Couche VIII link the geologic layer with Couche 4 at Roc de Combe. At that site, a rather late Aurignacian industry, the Aurignacian IV facies, was present in Couche 4 (Bordes and Labrot 1967; Laville 1975: 221-34). Chronostratigraphic contemporaneity between the Aurignacian IV of Roc de Combe and the Typical Aurignacian of Le Flageolet is somewhat surprising because the two facies were thought to be termporally distinct. The implications of Rigaud's suggestions are provocative, but the resolution of this problem must await the final sedimentology report.

In 1974, it was noted for the first time that Couche VIII contained more than a single archaeological occupation. But, given the location of the materials, i.e., between the massive rocks of Couche X, the division of levels was difficult to follow in some parts of the site. However, since all finds were mapped and recorded in three dimensions, vertical projections could be made from the data. These projections showed clearly that two

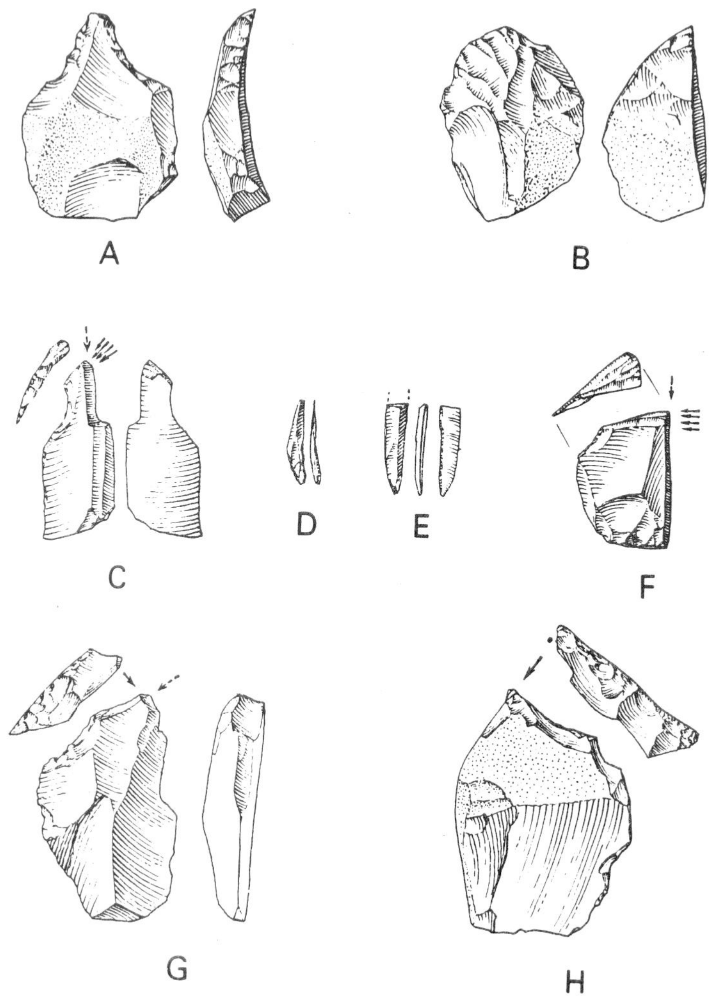

Figure 43. Retouched Stone Tools from Le Flageolet Couche VIII. The assemblage contains carinated and shouldered scrapers (A, B); busqued burins (C); Dufour bladelets (D, E); Carinated burins (F); and other burin forms (G, H). Drawings are two-thirds natural size. All illustrations from Rigaud (1982) and drawn by J-G. Marcillaud.

occupations were present in Couche VIII over the entire site. Materials that had already been excavated were separated based on projections, and excavations in the Couche after 1979 kept the surfaces distinct. The two archaeological levels were designated level VIII-1 (the upper surface) and level VIII-2 (the lower surface). The bulk of excavation in these two occupations occurred during the 1980 season, and the division between units was easily maintained.

After complete excavation, the level VIII-1 assemblage contains 214 tools, and level VIII-2 has an assemblage of 303 tools. In addition to the classic "tool types" composing these numbers, quantities of visibly utilized but unretouched lithics were recovered from both levels.

Using the type list developed by de Sonneville-Bordes and Perrot (1955-56) and later modified by a group of paleolithic archaeologists, assemblage composition can be compared between Couche VIII as discussed by Rigaud in 1980 and the two refined stratigraphic units. Figure 44 represents a comparison among the three units and shows that the two finer levels differ only slightly in typological terms from each other and from the Couche VIII profile. The magnitude of difference is not great enough for distinct assemblage attributions to be proposed. All are Typical Aurignacian.

More than 25% of the level VIII-1 assemblage is made up of endscrapers. Burins are not rare, however, and dihedral burins dominate within this class. Caminade endscrapers and Dufour bladelets are present. Busqued burins make up an important part of the industry (2.78%). Over 7% of the level VIII-1 assemblage is composed of carinated and shouldered endscrapers, contributing to the very Aurignacian character of the collection.

While similar to level VIII-1, the level VIII-2 assemblage shows some important differences. Again, over 25% of the industry is made up of endscrapers, and burins are dominated by dihedral forms. Busqued burins are present, but in reduced numbers in comparison with level VIII-2 (IBb = 0.99). Tiny Caminade endscrapers (see below) make up a substantial proportion of the scraper population; when this form is removed from the type counts, the relative contribution to the assemblage of endscrapers and burins is very close (a rare condition in Aurignacian assemblages).

Couche IX is only partially excavated at the time of this writing, but it has proved to be a dense archaeological surface yielding 532 tools so far. Major differences between level IX and the later levels lie in higher frequencies for certain typically Aurignacian types, e.g., carinated and shouldered endscrapers, Caminade scrapers, and carinated burins. A large

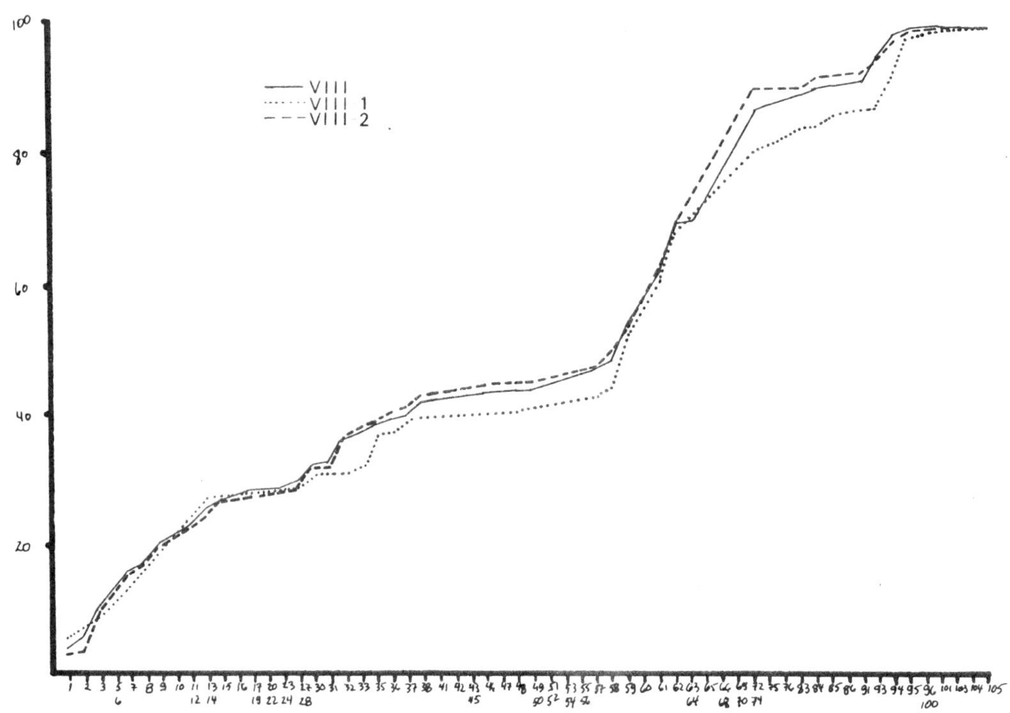

Figure 44. Typological Comparison between Couche VIII As a Single Assemblage, Level VIII-1, and Level VIII-2.

proportion of the level IX assemblage is composed of endscrapers, including some round and eventail forms associated with the early Typical Aurignacian. Caminades are very abundant (the 62 examples so far recovered from level IX represent over 80% of all known Caminades), and they thus compose a substantial proportion of the total endscraper population. If these enigmatic items are removed from the endscraper counts, the proportions of burins and endscrapers in the level IX assemblage are equal (20.68% vs. 19.0%). Again, busqued burins are present as are Dufour bladelets. Burins are mostly of the dihedral variety. In level IX, certain large blades show the invasive and scalar retouch called "Aurignacian" for its association with the earliest manifestations of this assemblage tradition.

AURIGNACIAN ASSEMBLAGE CLASSIFICATION

Based on similarities among Aurignacian assemblages from Roc de Combe, Caminade, and the industries from Le Flageolet I Couche VIII, Rigaud assigns both level VIII-1 and level VIII-2 to the Aurignacian I typological facies. Rigaud's tentative chronostratigraphic correlation between Couche 4 at Roc de Combe (containing an Aurignacian IV industry) and the presumably earlier Aurignacian I of level VIII-1 is provocative. The presence of Caminade scrapers in the Le Flageolet Aurignacian assemblages relates these to the Aurignacian II industries of the type site Caminade.

Characteristics of the level IX assemblage are similar to the Aurignacian III industry from La Ferrassie (Delporte 1977). This similarity creates further problems for general Aurignacian assemblage systematics (Rigaud, op cit.: 19). If Rigaud's assemblage assignments hold up after completed excavation of level IX, then the chronological relevance of traditional Typical Aurignacian facies must be reassessed. As for Upper Periogordian systematics, Le Flageolet seems determined to undermine conventional Aurignacian chronology.

Although Caminade scrapers appear in the later Typical Aurignacian layers at the type site, these have their most frequent occurrence in the earliest layer at Le Flageolet. The relative proportion of this type in the assemblage diminishes through the sequence, but Caminades are present in all Aurignacian levels. None of the Le Flageolet I surfaces can be assigned to the Aurignacian II facies on typological grounds. Yet it was with this facies that the Caminade tool type associated at the type site. Based on the data from Le Flageolet, Rigaud has argued that this tool type may have considerable time depth and is not a useful type-fossil (1982).

A fourth Aurignacian level is present in Le Flageolet I at and under the Couche X rocks. This level XI has been very partially excavated. The nature of the industry is unclear, since there are very few artifacts recovered at this point in time. Certain characteristics, most notably scalar retouch, would suggest an Early Aurignacian ascription. However, the sample is too small to make any definitive statements now, and level XI is not considered further in this work.

Despite certain differences, the Aurignacian industries from Le Flageolet I fall within the range of assemblage variation encompassed by the term "Typical Aurignacian." Problems in conventional systematics are raised by the Le Flageolet data, e.g., an apparant inversion of Aurignacian I levels above the assumedly later Aurginacian III. However, Rigaud suggests that many of the differences among assemblages heretofore assigned temporal significance may have other causes (variation in site use, for example) and that Le Flageolet I may represent a different kind of site that those in the Vezere Valley where traditional chronologies were developed (1980: 21).

RAW MATERIAL USE IN LEVEL VIII

A different kind of picture of the Le Flageolet Aurignacian emerges by considering raw material use during the deposition of Couche VIII. The archaeological assemblage considered is the combined level VIII-1 and level VIII-2 industries. Raw material use in Couche VIII has been discussed in a previous paper (Simek and Leslie 1983) so only a brief review will be presented here.

Variation in the cherts used by Le Flageolet occupants is great. Yet many people working with the materials have commented on the apparant selection of certain cherts over others for the production of certain tools. These "selected" cherts are often of high quality, but their sources lie many tens, even hundreds, of kilometers from the site. Paul Leslie and I decided to test the proposition of chert selection statistically to determine if what was an impression had empirical validity. We began by positing a model for on-site tool production using different kinds of chert. This model was based on lithic waste materials and the proportion of the debris assemblage comprised of four different cherts. We took this model to represent the actual working of chert during the deposition of Couche VIII.

Using a method that partitions the chi-square statistic, we tested the hypothesis that burins, utilized pieces, and other tools in Couche VIII had similar proportions of chert types. We found that certain tools, burins in particular, were made from cherts available only to the distant west of Le Flageolet. In fact, all tools, excepting utilized pieces, had a greater proportion made from "exotic" cherts than the debris model predicted. Debris was made up predominantly of locally available cherts. We interpreted these results as indicating curation, so that tools deposited in the shelter had not been made there and tools made in the shelter had been exported.

Raw material use at Le Flageolet was certainly more complex than can be described by the concept "curation". Still, the general picture of mobility and fluidity indicated by this test is supported in the spatial analyses that follow. Another bit of evidence supporting this view involves refitting debitage and stone tools. For both levels in Couche VIII, attempts were made to refit burins and the spalls removed from them to form the bit. Not a single join was found. This supports the notion that burins deposited at Le Flageolet were made elsewhere, and burins made at the site are no longer there.

Evidence from raw material studies expands on inferences suggested by the ecological data discussed above. It appears that the Aurignacian occupants of Le Flageolet brought a set of tools with them when they occupied the site. The raw materials from which these tools were made may inform on the direction of travel (i.e., from the west), although we must know more about the relative "values" of different chert types before the possibility of trade can be ruled out. While at Le Flageolet, these people intensively utilized resources from close to the site, including chert available on the plateau above. They used up some of the tools they brought with them during the local exploitation at Le Flageolet and made new tools which they conserved upon departure.

ATTRIBUTE ANALYSES OF CERTAIN AURIGNACIAN LITHIC ARTIFACTS

Another method used to compare assemblages is the analysis of attributes. For the present study, only materials from level VIII-1 and level VIII-2 are considered because the other Aurignacian assemblages are not yet completely excavated. The two Couche VIII levels are so similar in typological terms that it seems fruitful to see just how similar they are in other ways. Attribute comparisons would seem to be a useful way to approach fine-scale morphological variation since they concentrate on characteristics of objects rather than on the objects themselves. It should be noted that variability in attributes between the two levels might be due to a multiplicity of factors, e.g., functional variability, stylistic differences, technological variation, etc. In this study, no causal inferences will be made in the attribute comparisons. It is simply the measurement of differences between level VIII-1 and level VIII-2 that is of concern.

Core Morphology

Cores recovered from the excavation of levels VIII-1 and VIII-2 are classified according to five techno-morphological groups. These classes are defined based on the pattern of debitage removel. The five core classes are:

1) Unipolar Prismatic. These are cores on which the majority of scars are twice as long as they are wide and were produced by removals from a single striking platform.

2) Bipolar Prismatic. These are cores on which the majority of scars are twice as long as they are wide and were produced by removals from two striking platforms, usually opposed at the ends of the core's long axis.

3) Pyramidal. These are cores on which the majority of scars are less that twice as long as they are wide and were produced by removals from a single striking platform. The platform is usually round, and its entire circumference was used to remove flakes. A cross section of the core is triangular.

4) Globular. These are cores on which the majority of scars are less than twice as long as they are wide and were produced by removals from many striking platforms. A cross section of the core is round.

5) Amorphous. These are cores with scars of varying sizes that were produced by removals from a variety of platforms. No geometric form can be used to characterize the shape of these cores.

Table 20 compares the frequencies of each of these core classes between level VIII-1 and level VIII-2. A chi-square test for homogeneity in the Table (Kempthorne 1969), performed under the null hypothesis of no difference between the collections, cannot reject that possibility. There is no significant difference between the levels in how and to what extent cores were reduced prior to abandonment. Differences might have been anticipated if resource availability was greater for one of the occupations, allowing reduction to cease after the best blades had been removed from the core (leaving numbers of prismatic and bipolar prismatic forms). Different reduction technologies might also have resulted in variation in core form between levels VIII-1 and VIII-2. That such differences are not present suggests

TABLE 20.

Frequencies of five core classes for Le Flageolet I levels VIII-1 and VIII-2. A chi-squared test for homogeneity between the two levels cannot reject a null hypothesis of no difference (N = 25 in both cases).

CORE CLASS	VIII-1	VIII-2
Unipolar prismatic	3	4
Bipolar prismatic	1	5
Pyramidal	4	1
Globular	11	5
Amorphous	6	10

that the same reduction technology was used and that access to lithic resources was equivalent. The high incidence of globular and amorphous cores in both assemblages suggests that reduction continued as far as possible in many cases--to the point where only flakes could be removed. It is, of course, possible that still-useful cores were exported from the site at abandonment (and this is consistent with the locally intensive land use pattern at Le Flageolet posited earlier), but it would appear that any such curation was practiced in a similar way in both occupations.

Core Weight

To examine if abandoned cores differ between level VIII-1 and level VIII-2 in size, the mean weight of cores in each level is computed and standard deviations calculated. For level VIII-1, the average weight is 147 gms (s = 199.486). For level VIII-2, the mean is 118.667 gms (s = 191.761). There are 25 cores recorded for VIII-1 and 25 for VIII-2. A two-tailed *t*-test was carried out to compare these values (Blalock 1972). The null hypothesis tested is that the means are equal and could derive from the same population. In this test, H0 must be accepted; there is no difference in core weight between the two levels.

Core Raw Material

A second set of comparisons is made to determine if there are differences between level VIII-1 and level VIII-2 in the materials from which cores are made. Three different kinds of local raw material are distinguished: (1) a grey, grainy chert deriving from Santonian limestones and found today some distance to the north of Le Flageolet, (2) a black, waxy, fine-grained chert found near Les Eyzies to the northwest, and (3) a blue-black, fine-grained chert that occurs in small nodules on the plateau just above the site. (It should be pointed out that the study of chert sources is being carried out over much of the Perigord region by a team of archaeologists of which the author is *not* a member. The categories distinguished here are based on discussion with these scholars, but they will surely be subject to modification as research proceeds.) *T*-tests are again used to examine differences between level VIII-1 and level VIII-2 in core weights within these three chert classes. Two sets of tests are performed. One set evaluates the null hypothesis that there are no differences between levels in core weight within a material class. The second set tests for differences in core weight between material classes within a single archaeological level. The null hypothesis cannot be rejected in any test. Discarded cores appear to be of similar size regardless of the material involved, and cores are of the same sizes in both levels.

Summary of Core Attribute Comparisons

Chert selection, reduction, and discard seems to have been the same in both level VIII-1 and level VIII-2. Debitage and lithic debris are composed primarily of locally available cherts in both levels, while a higher proportion of tools are made from exotic cherts than would be expected from the waste composition. Some artifacts show cortical surfaces with scars from frost cracking, indicating that nodules were obtained from the plateau, where this kind of weathering affects materials, in addition to more conventional bedded sources. This pattern is evident in both levels.

For both level VIII-1 and level VIII-2, a blade technology is indicated by the tools, while cores are mostly reduced to globular and amorphous forms that could only yield flakes prior to abandonment. In both occupations, cores were discarded when of more or less uniform size, although variation in size does exist that may relate to the quality of material. Core size is indistinguishable between levels. There are no significant differences between levels in terms of the raw material discarded as cores. It seems that all chert was reduced as far as possible, either to the point where no more removals could be obtained from a core or until internal flaws made the core useless. The pattern of chert utilization indicated here lends further support to the idea that Le Flageolet Couche VIII represents at least two short term, locally-intensive occupations.

Tool Weight

As for cores, the weights of various general artifact classes are compared between level VIII-1 and level VIII-2. Weight was selected as a relevant variable here since it is a general indicator of artifact size. For many tasks, the strength of a tool is important both in its function and in its production. Moreover, the shape of retouched edges (thickness, angle, etc.) are conditioned by the size and shape of the blank selected for transformation. Simply put, a tool designed to carry out a given task was probably manufactured on a flake or blade of the appropriate size and/or heft.

Seven artifact categories are considered, including endscrapers, Aurignacian pieces, burins, truncations, miscellaneous retouched pieces, notched and denticulated pieces, and pieces showing obvious utilization damage on their edges. These categories represent general technological classes that subsume certain amounts of the formal variation defined in the de Sonneville-Bordes and Perrot types. The classes used here are, in fact, sets of the classic types. Table 21 relates the classes used to the de Sonneville-Bordes and Perrot typology.

Classic types are collapsed into these general classes because the nature of variability expressed in the traditional typology is unclear. More general categories have the advantage of strict technological definition. For example, the class "burins" comprises all artifacts where a transverse blow was struck to the blank. Preparation of the blank and shape of the resulting edge are ignored, although these criteria are often used in the traditional typology.

Mean artifact weight and the standard deviation around the mean are calculated for each class by archaeological level. _T_-tests are carried out to test the hypothesis of equivalent means between data from level VIII-1 and level VIII-2. No differences are present between the two occupations. _T_ values are, in fact, infintesimal, and significance for the statistic is never obtained despite the many degrees of freedom accompanying each test.

Again, similarilty between level VIII-1 and level VIII-2 is indicated. Tool weight is probably related to raw material reduction technology as well as to anticipated tool use. We have already seen that there are no differences between the levels in this dimension.

Attributes of Endscrapers

Eight attributes of endscrapers are measured to compare between archaeological levels. These include: the width of the endscraper bit, bit length, the radius of the endscraper front, the arc of the front, the angle of the endscraper bit from the edge to the highest point of retouch, the angle of the bit just at the juncture with the ventral face, the thickness of the bit, and the angle of symmetry of the scraper front. These attributes have been used by many archaeologists (e.g., Wilmsen 1970); their characteristics and measurement are considered in detail by Movius and his co-workers at the Abri Pataud (Movius, _et al._ 1968; Movius and Brooks 1971; Sackett 1966). While Movius and his colleagues collapse measured attributes into nominal categories in many cases, continuous measurements are used here because of certain test statistic requirements.

_T_-tests comparing endscraper attributes between level VIII-1 and level VIII-2 assemblages again fail to reject the null hypothesis of no differences. Certain of the attributes may relate to blank size (endscraper front

TABLE 21.

Relationships of the general artifact classes used for Le Flageolet I spatial analyses to the new Upper Paleolithic type list (cf. Bordes 1978; Rigaud 1982). Note that some type list numbers are repeated since they are composite tools having more than one edge (e.g., 17).

GENERAL CLASS	TYPE NUMBER
Endscrapers	1-6, 8, 9, 17
Burins	17, 19, 30-46
Aurignacian Pieces	11-16, 33, 35
Truncated Pieces	57-59
Notched Pieces	72-74
Retouched Pieces	61-63, 75
Utilized Pieces	no reference

thickness, for example), and this is demonstrably equivalent between the occupations already. Other attributes do not obviously relate to blank form (e.g., front symmetry, arc, radius, and edge angle). Yet these, too, are indistinguishable between levels. Some of these attributes may have their primary sources of variation in functional or stylistic factors. Symmetry and radius, both of which are produced by intentional alteration of blank form by retouch, might be such characteristics. There is no difference between level VIII-1 and level VIII-2 in these attributes.

The angle of the endscraper edge may be related to the use of the tool. To examine endscraper use, two angles were measured at the bit (see above). The first endscraper angle was measured from the edge to the top of the retouch, and might best be described as the "face angle" of the endscraper, since it represents the angle formed by manufacture of the tool. The second measurement was taken right at the endscraper bit and is a more-exact "edge angle". The edge angle may conform to the face angle or diverge from it as use altered the bit. Figure 45 (A and B) illustrates the different distributions of these two angles within each Couche VIII level. Edge angles are consistently steeper than face angles and are often formed by obvious damage of the bit. That no difference exists between level VIII-1 and level VIII-2 on either attribute suggests that endscrapers were used in similar fashion in both occupations. In short, the pattern of what appears to be use damage on endscrapers is the same in the two levels.

Attributes of Burins

Five attributes are measured for burins. Again, these were selected from the list proposed by Movius, *et al.* (1968). Measured burin attributes include: the length of the burin bit, the depth of the bit (since many burins are formed by multiple blows and have curved edges), the angle of the bit from the long axis of the blank, the angle of the edges forming the bit itself, and the angle of the bit from the ventral face of the blank. One nominal attribute was observed; the position of the bit with respect to the blank was classified into one of five states. These states involve dividing the burin lengthwise into five equal sections, numbered from one to five from left to right. Position is recorded by the corresponding section number. Lateral burins, thus, have attribute values of one or five (depending on which edge they occupy); angle burins (fr. *dejete*) usually get values of two or four; and axial burins receive an attribute value of three.

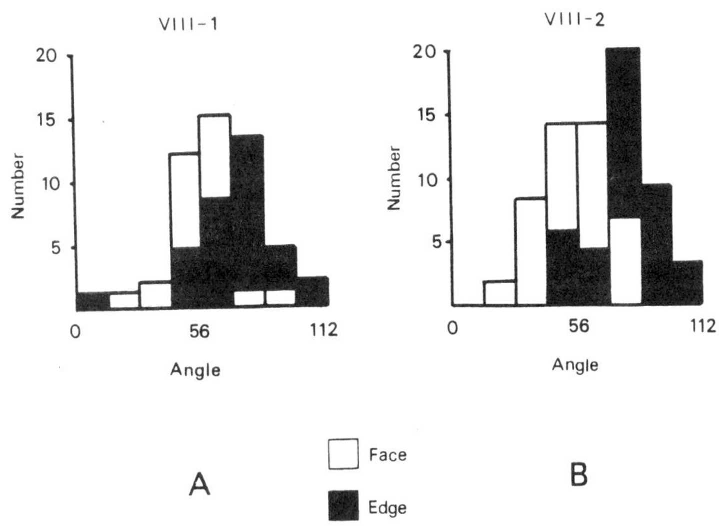

Figure 45. Comparison of Endscraper Face Angles to Edge Angles in Le Flageolet I Levels VIII-1 and VIII-2.

T-tests again reveal no significant differences between the levels for any of the continuous variables measured. Comparison between level VIII-1 and level VIII-2 for burin position is accomplished using a chi-square test for homogeneity as was discussed earlier. The resulting chi-square value of 4.29 is not significant. The null hypothesis--that there is no difference between the two levels in frequencies of burins by position--cannot be rejected. Again, levels VIII-1 and VIII-2 are similar in the forms of burins they contain.

There is almost total coincidence in the assemblages from levels VIII-1 and VIII-2, extending from overall assemblage typology, through raw material use and lithic reduction technology, to fine-scale characteristics of the dominant stone tool forms. This similarity probably involves both stylistic and functional aspects of artifacts, since no attempt was made here to distinguish between these dimensions of variability. Only one minor difference is present between the levels, and this pertains to an ubiquitous yet poorly understood category of artifacts.

Attributes of Pieces Showing Utilization Edge Damage

A final set of observations was made on artifacts that, perhaps, reflect tool use more obviously than many of the extensively retouched classes. This category consists of flakes and blades that show fine flaking or crushing of an edge produced directly by utilization. Members of this class may or may not have other, intentional retouch present. Often, damage is observed on otherwise unmodified lithics. Pieces classed as utilized pieces may not be the only unmodified tools in the Le Flageolet assemblages, as microscopic wear traces have often been found on objects showing no damage when viewed with the naked eye (e.g., Keeley 1980; Odell and Odell-Vereecken 1980). However, for the present study, all lithics were observed without magnification, and microscopic wear traces are not considered.

Two aspects of visible damage are recorded. The position of the utilization on the piece is determined by a series of eight standardized locations. Figure 46 illustrates how these locations are measured. The angle of the damaged edge was also measured. Attributes of utilization were measured both for otherwise-unmodified and for retouched pieces showing areas of damage away from the modified portions.

Figure 47 compares frequencies of utilization by location for unmodified and modified pieces for level VIII-1. The same comparison for level VIII-2 is shown in Figure 48. In both of these graphs, only unbroken pieces are considered since fractured objects distort the

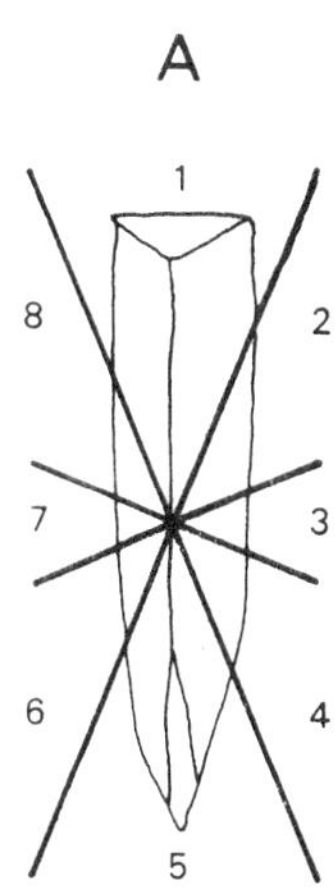

Figure 46. Radial System Used to Locate Edge Damage for Le Flageolet Artifacts. The system is shown for a blade (A) and a flake (B).

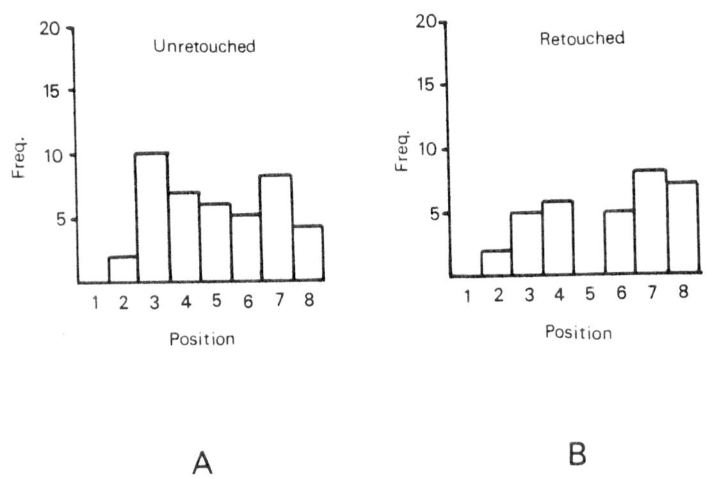

Figure 47. Frequencies of Visible Edge Damage for Level VIII-1 Unmodified (A) and Modified (B) Lithics by Position.

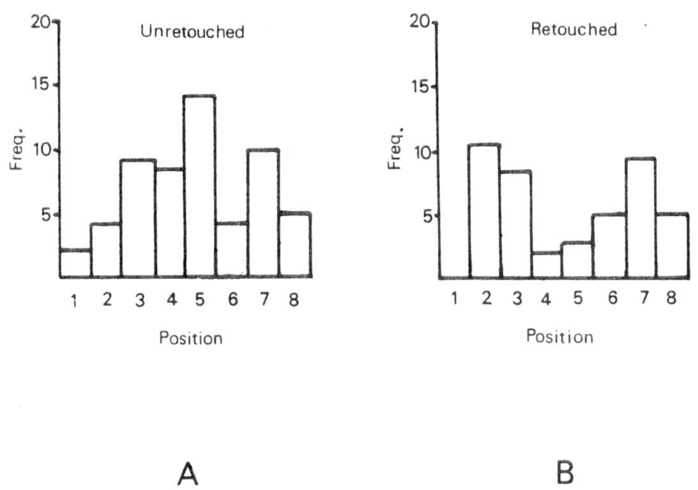

Figure 48. Frequencies of Visible Edge Damage for Level VIII-2 Unmodified (A) and Modified (B) Lithics by Position.

positional frequencies.

Because retouch tends to occur at the end of blanks (e.g., for endscrapers and burins) damage tends to be located along the lateral edges of retouched pieces. Even when unretouched, artifacts from level VIII-1 conform to this pattern of lateral use. For unmodified objects in level VIII-2, damage is more frequent at the distal ends of blanks. This difference between levels is interesting because it suggests some divergence in the use of unretouched lithics. However, this is the only attribute difference detected between the levels and indicates, at best, only slight functional variation.

Figure 49 compares the edge angles at utilization between level VIII-1 and level VIII-2 by position. The proximal or platform end (position 1) is not used very much in either level. The mean angle is similar for both levels. All value ranges overlap, and there are no modalities in angle distributions within positions. Again, no differences between the two Couche VIII occupations are indicated.

Summary of Attribute Comparisons

Several sets of attribute observations were compared between the two assemblages from Couche VIII. The weights of objects were compared between levels by artifact classes, but no significant differences were found. A series of endscraper characteristics were compared between the levels, but the levels were identical. No differences were detected in burins. The only difference found between level VIII-1 and level VIII-2 was in the more frequent appearance of damage at the ends of pieces in level VIII-2. Other aspects of utilized pieces were the same between the levels.

CONCLUSIONS

The picture of inter-assemblage variability obtained from these analyses is one of remarkable similarity in all aspects between Le Flageolet level VIII-1 and level VIII-2. Raw material procurement and reduction technology, stone tool design and manufacture, and tool use; all seem to be essentially identical between the two occupations. Other attribute comparisons might yield significant differences between the two levels. Using the present set, which involves attributes measuring the use of lithic resources and aspects of the working edges of tools, the two levels are the same.

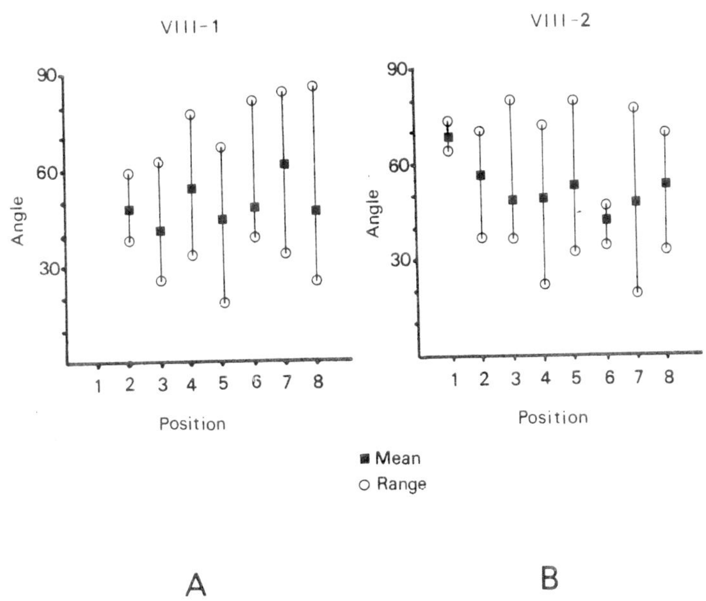

Figure 49. Ranges of Damaged Edge Angles by Position for Levels VIII-1 (A) and VIII-2 (B).

While it would appear that level VIII-1 and level VIII-2 are the same in terms of resource use and stone tool function, they might well vary in other ways. For example, the size of the occupying groups, the duration of occupation, or the composition of occupying parties might have been different even though _what they did_ at the site (i.e., site function) was the same. These kinds of differences will be manifested in characteristics of the sites other than artifact form, like spatial distributions of objects or features. It is to these topics that we now turn.

CHAPTER 6

SPATIAL CONTEXT AT LE FLAGEOLET

EXCAVATION TECHNIQUES

The recovery techniques employed by Rigaud at Le Flageolet I were designed to locate and record every artifactual object in three dimensional space. This includes all stone materials introduced into the shelter, i.e., not part of the natural fill process or modified by human activity. All bones, whole and fragmentary, were also mapped and located by Cartesian coordinates. Any characteristics of the sedimentary deposits of archaeological interest (stains, rock alignments, etc.) were mapped and discussed in fieldnotes. Photographs documenting the excavations were systematically made. During all work, particular attention was paid to stratigraphic problems, both natural and cultural, and control over the archaeological stratigraphy was maintained through vertical projections. A data set allowing the almost complete reconstruction of the site resulted from these excavation methods.

Despite recovery goals, no system is perfect, and some artifacts were missed as their sedimentary matrix is removed. Therefore, all sediments were screened after removal, and contents were saved by archaeological level and meter square. Materials recovered during screening were provenienced by grid units (see Figure 42 in Chapter 5). For the most part, objects found in screens are under two centimeters in maximum dimension but include microlithic tools (e.g., Caminade scrapers). Still, every excavator tried hard to find all artifacts *in situ*. Thus, many objects under two centimeters in maximum size were found and mapped; to be consistent with the bulk of materials this size, small finds were not given coordinates even if they were mapped. Data for the small size class (< 2 cm) are in the form of frequency counts and weights per meter square grid unit. Nearly all objects larger than 2 cm have point provenience.

CONTEXTUAL ELEMENTS

The *k*-means based approach advocated by Kintigh and Ammerman (1982) requires that the context of spatial distributions be integrated into the analysis. Therefore, certain classes of data must be discussed as a background to the formal analyses which follow. These include the nature of the space available to the occupants of Le

Flageolet at various times during the paleolithic (i.e., the paleotopography). Features, which served as the foci for certain kinds of deposition at Pincevent 36, must also be described in reference to space use and organization at Le Flageolet. In discussing contextual characteristics, the data collected by meters are examined since these may show certain patterns in their distributions that can be related to the point data.

Paleotopography

For each of the Couche VIII Aurignacian occupations, the massive roof collapse labelled Couche X influenced the nature of deposition and the resulting artifact distributions. Figures 50 and 51 illustrate the plan of blocks for level VIII-1 and level VIII-2. The blocks on these maps were, in many cases, at least partially buried in sediments at the level where occupation debris was found. Thus, the surface of each "level" varies with respect to block surfaces even where the same rock is present on two maps. The northern sector of the site, for example, contains large, flat rocks whose surfaces are only slightly more elevated that the surrounding matrix. Relief is not dramatic. In the central portions of the site, relief is very great. Sediments dating to the Aurignacian occupations occur deep within cracks between rocks. Moreover, the surface altitudes of the central rocks are also variable. The central sector of the shelter must have been very uncomfortble to move across (as it is today). In the south, another area of flat rock surfaces dominates the topography. Between these large rocks are deep crevices, but these cracks tend to be wider than in the north and center of the shelter. Archaeological materials dating to the Aurignacian are found in these cracks. The altitude of rock surfaces in the southern sector is approximately one meter above the northern sector surfaces.

These conditions were present during all Aurignacian occupations at Le Flageolet. The depth of crevices between rocks diminished from level IX to level VIII-1. It is clear from the rock maps that periodic roof collapse, of lesser magnitude than that producing Couche X, occurred (hopefully!) between occupations. Thus, the surface topography varied from one occupation to the next.

In general, the shelter interior available to Aurignacian occupants of Le Flageolet involved two relatively flat areas (considering the block surfaces) separated by an uneven jumble of odd-shaped boulders. The southern sector, from the 6 line to the 11 line on the sagittal grid axis, was always about a meter higher than the northern flat surface. Crevices between boulders were, for the most part, narrow and deep, especially during the level IX occupation. As sediments and refuse

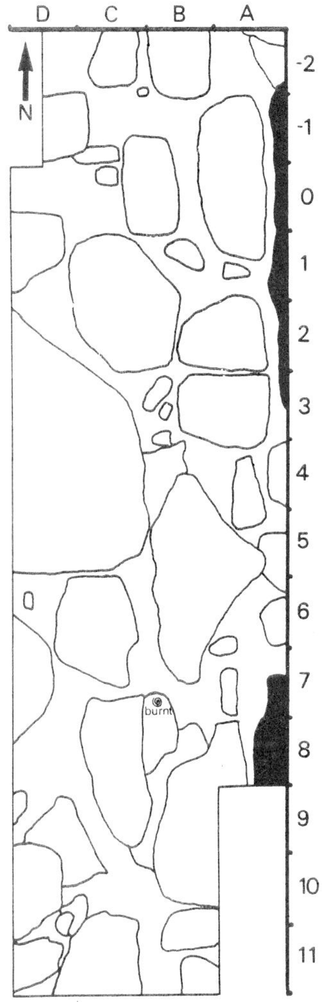

Figure 50. Plan of Blocks Present during the Deposition of Le Flageolet I Level VIII-1.

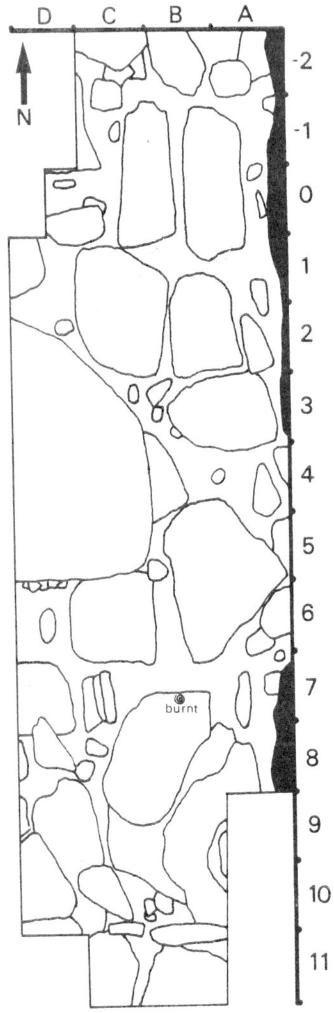

Figure 51. Plan of Blocks Present during the Deposition of Le Flageolet I Level VIII-2.

filled the cracks, relief diminished in some areas. However, crevices were present below the rocks throughout the period of Aurignacian site use. What cultural features are known were constructed on the surfaces of the blocks.

Cultural Features, Level VIII-1

A single hearth was found in level VIII-1, located in the northern sector of the surface. This hearth, centered in grid units A1 and B1, will be referred to below as the "A1 hearth". The feature is composed of burnt and unburnt igneous cobbles, some dark staining, and abundant burnt bone. However, the spatial structure of these elements is diffuse, and they are not numerous. Other cobbles and fragments in level VIII-1 are distributed around the surface with no apparent pattern.

A *k*-means analysis of all cobbles and fragments over 2 cm in maximum dimension shows optimal cluster solutions when three and five concentrations are defined (Figure 52). When three clusters are considered, ten pieces are in the A1 hearth, five are located to the north near the shelter wall, and six are present in the southern sector (Figure 53). At the five cluster level (Figure 54), the A1 area is divided into two clusters with one centered in grid unit B0. These hearth clusters again contain ten objects. Five pieces are concentrated in unit A-2. Two clusters of three cobbles each are now defined in the southern sector. The only real accumulation of cobbles is in and near the A1 hearth.

Figure 55 illustrates the level VIII-1 hearth. The main concentration of cobbles occurs in A1, while other objects are scattered into units B1 and B0. In the five cluster solution, *k*-means defines the B1/B0 scatter as a separate cluster. *It* is interesting to note that the only charcoal present in this area is in B1. However, it is not uncommon to find carbonized materials outside of hearth features where they were deposited (assumedly) during maintenance, i.e., cleaning or sweeping, of the hearth.

The distribution of igneous cobble fragments less that 2 cm in maximum dimension is illustrated as a density map in Figure 56. A lack of correspondance is evident between the distribution of larger cobbles and fragments (centered in A1) and these smaller pieces. Small cobble fragments have their greatest frequency in southern sector unit B6. While none of the *k*-means clusters for cobbles are centered in this *unit*, there are nearby clusters in both the three and five cluster solutions. The northern part of the south sector seems to be a focus for distributions of large and small cobble fragments. No systematic pattern in these materials is evident that

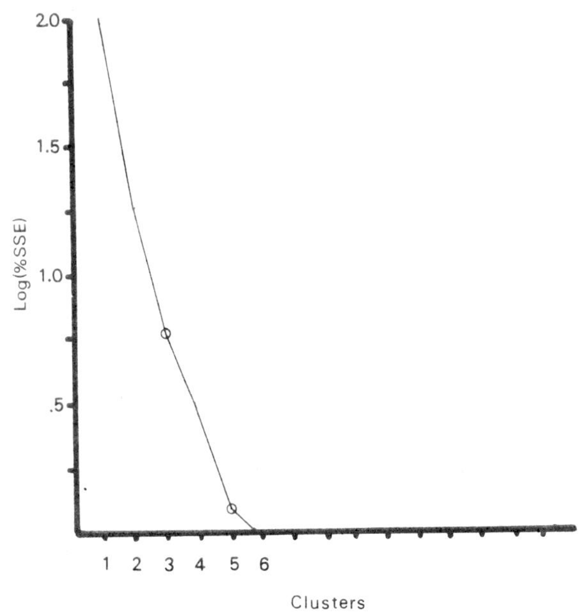

Figure 52. Plot of log(%SSE) Statistics for Cobbles from Le Flageolet I Level VIII-1 Produced by K-means Cluster Analysis to Fifteen Clusters.

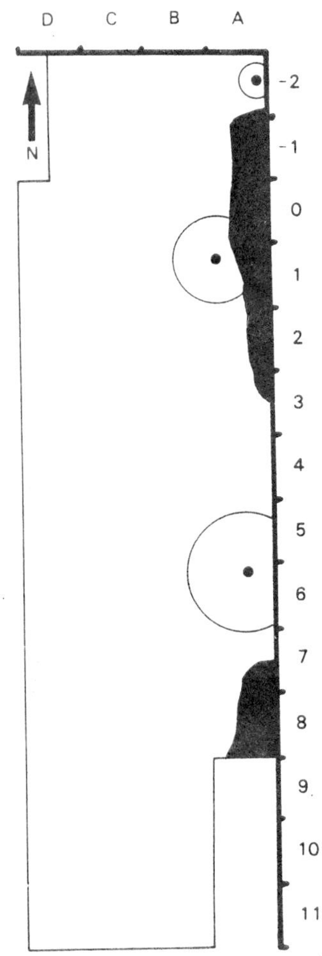

Figure 53. Map of Three Cluster K-means Solution for Le Flageolet I Level VIII-1 Cobble Distribution. Dots mark cluster centroids; circles are RMS radii for clusters.

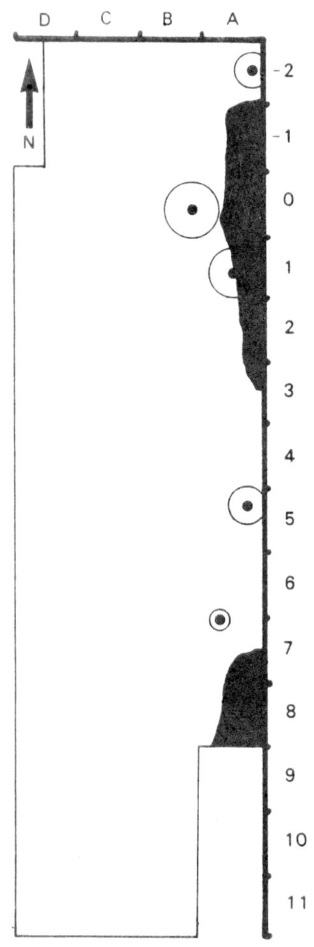

Figure 54. Map of Five Cluster K-means Solution for Le Flageolet I Level VIII-1 Cobble Distribution. Dots mark cluster centroids; circles are RMS radii for clusters.

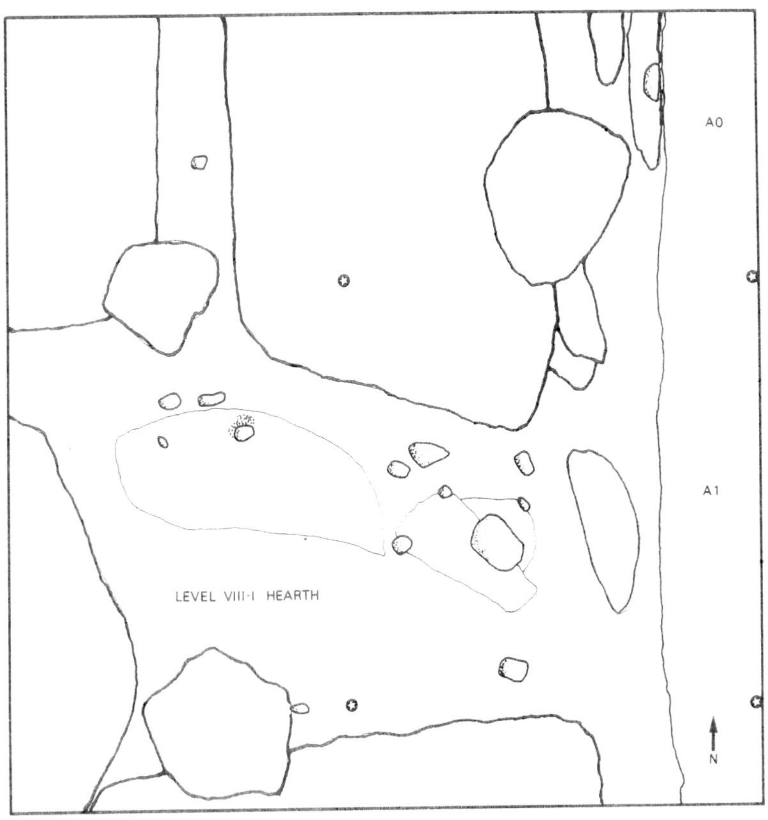

Figure 55. Plan of Hearth Feature for Le Flageolet I Level VIII-1.

might suggest the presence of a second hearth. Small fragments are also concentrated near the A1 hearth. However, here small fragments are concentrated in unit C0. A few small pieces occur between this unit and the hearth in A1, but the two areas are essentially separate. Other small fragments are scattered over the site surface, but these are not associated with clusters of larger cobbles.

A general spatial pattern is evident in those areas containing concentrations of both large and small cobbles. Larger pieces seem to be clustered to the east of small fragments (towards the rockshelter wall). This relationship cannot be explained by sorting during natural erosion for two reasons. First, the slope in Le Flageolet deposits follows the bedrock towards the central fault. In Le Flageolet I, the slope is from south to north, not east to west. Second, the large blocks limit movement in all directions, but the lengthiest crevices also trend from south to north. It seems that the pattern in cobbles is cultural rather than natural in origin and may reflect the cleaning of fire features.

As was stated earlier, the A1 hearth is very diffuse, and there is no evident organization in the feature elements. The A1 feature must have been small given the number of cobbles involved. If there was ever a feature near unit B6, it must have been even smaller considering the number of rocks and the total lack of visible evidence for feature construction.

Several classes of material suggest that larger cobbles mark the original position of the hearth and that the smaller fragments were removed from the feature. These include bone, dechet (lithic waste < 2 cm) and debitage (lithic waste > 2 cm). These classes have overlapping distributions near the A1 hearth. There are differences in their distributions over the entire surface, particularly in the southern sector. All of these classes are concentrated in unit C0 to the northwest of A1, and this accumulation may relate to the hearth area. Southern accumulations are different among the classes and are not related to any surviving feature.

Bone debris, although it was collected with point provenience, is examined here in terms of unit density because identification is not yet complete. Figure 57 shows the density map for this class. Bones seem to be concentrated in two places. The richest accumulation occurs in unit A-1 near the shelter wall, but a band of units trending east to west (including unit C0) contains quantities of bone debris.

Dechet shows a similar pattern to cobble fragments. Figure 58 is the density map for this class and illustrates the concentration of dechet in units C0 and D0. There are quantities of dechet in unit B1, suggesting a link with the A1 hearth. Only one southern unit, B6,

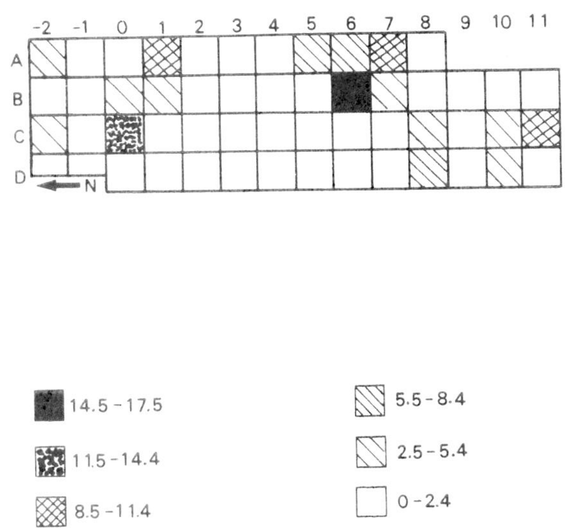

Figure 56. Density of Cobble Fragments (< 2cm) by Meter Square from Le Flageolet I Level VIII-1.

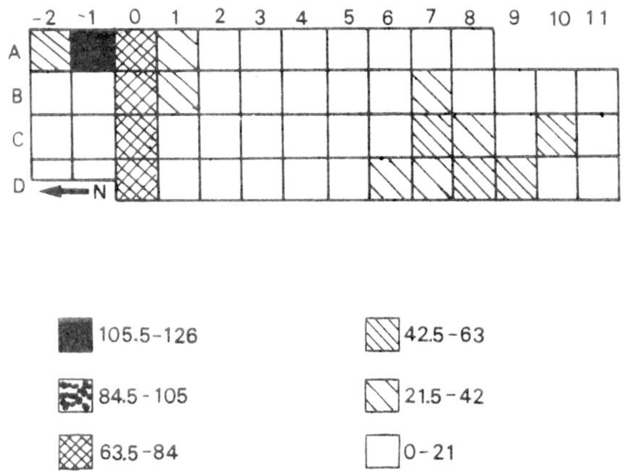

Figure 57. Density of Bone (> 2cm) by Meter Square from Le Flageolet I Level VIII-1.

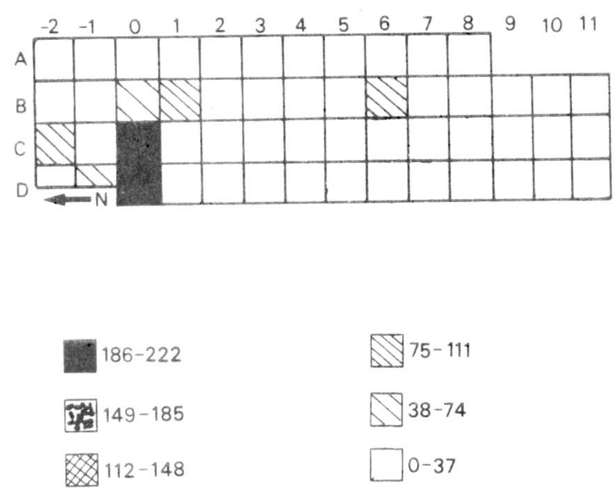

Figure 58. Density of Dechet (Lithic Materials < 2cm) by Meter Square from Le Flageolet I Level VIII-1.

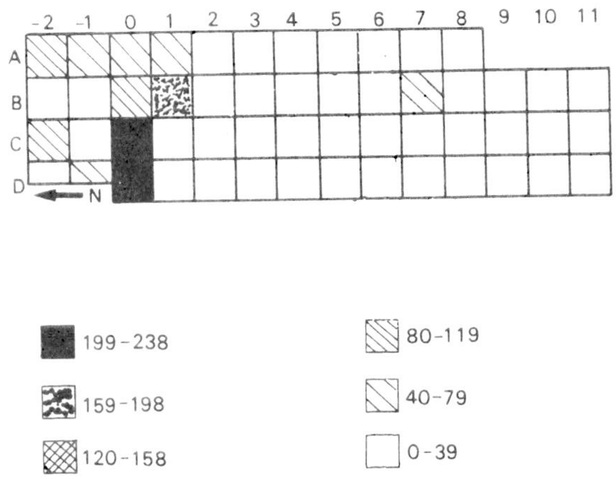

Figure 59. Density of Debitage (Lithic Materials > 2cm) by Meter Square from Le Flageolet I Level VIII-1.

contains many of these small lithic chips. Dechet concentrations may reflect the loci of stone tool production, as this is a frequent and ubiquitous by-product of that process. Thus, it would seem that most chert reduction occurred in proximity to the A1 hearth and waste materials were either directed or moved to refuse piles west of the feature itself.

The distribution of debitage supports this inference. The correspondance between debitage densities by meter square with dechet distributions is striking (Figure 59). Again, there is a concentration of materials in units C0/D0, and only B7 has a relatively high density in the southern sector. The spatial overlap between these two lithic size classes indicates that post-depositional sorting did not affect these accumulations. Coincidence among debitage, dechet, and small cobble fragments in units C0/D0 suggests that this crevice was a dump for the unwanted products of hearth area activities. The presence of both debitage and dechet concentrations in units B6 and B7 suggests that lithic reduction occurred near these southern sector units. The amount of lithic waste deposited in the south is much less than in northern hearth area accumulations.

Summary of VIII-1 Contextual Data

In summarizing the basic context within which spatial analysis of level VIII-1 artifact distributions will proceed, the following characteristics are relevant. A single hearth feature is present, centered in unit A1. Just to the north and west of this feature, accumulations of material that seem to reflect the "sweeping out" of hearth materials are scattered over flat block surfaces. The hearth itself is small and diffuse and suggests short term use. It is possible that a second feature was present somewhere in the southern sector of the site (near unit B6), but physical evidence for this feature is scant. Accumulations of bone, dechet, and debitage in units C0/D0 suggest that this space served for disposal of hearth activity refuse. Other concentrations of material near the shelter wall north of the hearth may reflect more-dispersed areas of discard. Artifacts are scattered over the southern sector of the site, but their relationships do not systematically overlap as in the north.

In terms of activities, stone tool production seems to have occurred frequently in the hearth context, and waste products were discarded into dumps located a slight distance away from the feature. Lithic reduction also occurred in the southern sector, but refuse is less abundant than in the north and has no evident feature association.

One bit of preliminary evidence supports the notion that a fire feature, now invisible, was present in the southern sector at some time. However, it is not possible to relate this feature to any Aurignacian occupation with certainty. The evidence comprises carbon signal readings by Electron Spin Resonance Spectroscopy (Robbins, *et al.*, 1978) for a series of samples taken from the surface of a southern sector rock. The area where these samples were removed is in unit B7 above the southern sector cobble concentrations (see Figure 50). Robbins' tests demonstrate that this block was, indeed, subjected to heat at some past time (Robbins, personal communication). There is also visible fire damage to the area. However, to which level(s) the feature is related is unclear. As will be seen, the characteristics of level VIII-1 in this area are also present in level VIII-2. It may be that both occupations, or neither, involved the use of a small fire in the B7 rock surface. It is not presently possible to make an accurate ascription.

Cultural Features, Level VIII-2

As in level VIII-1, a single hearth feature was uncovered in VIII-2. Again, the feature is located in the northern sector of the site and overlaps partially with the VIII-1 hearth. However, this feature is extensive and seems to have been made up of several parts. The parts may represent the actual hearth position, hearth sweepings, and refuse associated with the feature. Overall, the level VIII-2 hearth seems to have been more intensely used than its level VIII-1 counterpart.

Over 200 cobbles and fragments were entered into *k*-means cluster analysis, and optimal solutions at two and seven cluster configurations are indicated (Figure 60). When two clusters are considered (Figure 61), one centroid is located in grid unit B0 and the other in C7. The northern cluster has a point frequency of 184 while the southern cluster contains only 29 objects. At the seven cluster level (Figure 62), four clusters are located in the north, and three of these are very rich (with 52, 57, and 61 pieces). Three sparse cobble clusters are present in the southern sector, all with centroids in crevices around the B7 block.

Plan maps of the three rich northern concentrations were made during the excavation of level VIII-2. Figure 63 shows the one centered on unit A1. This concentration is composed of large pieces; very few small fragments are present. The sedimentary matrix of this A1 cluster does not contain the carbonized organic materials that might indicate *in situ* burning. A second rich cobble concentration, this one containing carbon and burnt limestone, is centered on grid unit B0. Tests of the sediments within this concentration indicate exposure to

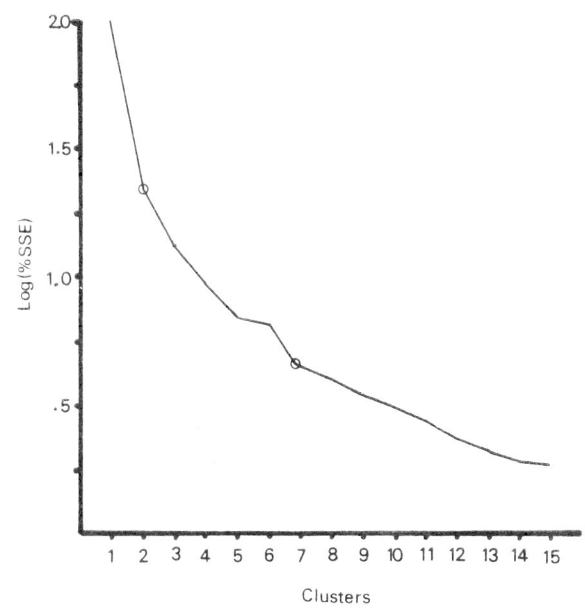

Figure 60. Plot of log(%SSE) Statistics for Cobbles from Le Flageolet I Level VIII-2 Produced by K-means Cluster Analysis to Fifteen Clusters.

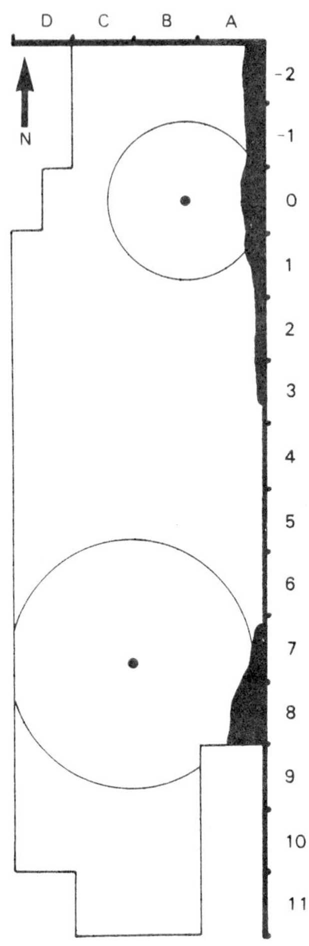

Figure 61. Map of Two Cluster K-means Solution for Le Flageolet I Level VIII-2 Cobble Distribution. Dots mark cluster centroids; circles are RMS radii for clusters.

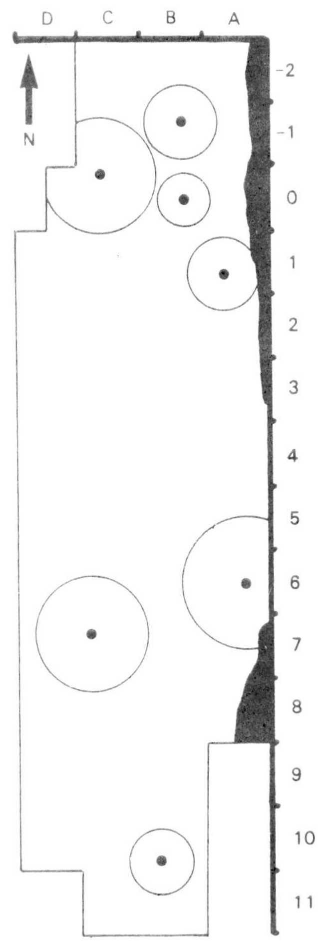

Figure 62. Map of Seven Cluster K-means Solution for Le Flageolet I Level VIII-2 Cobble Distribution. Dots mark cluster centroids; circles are RMS radii for clusters.

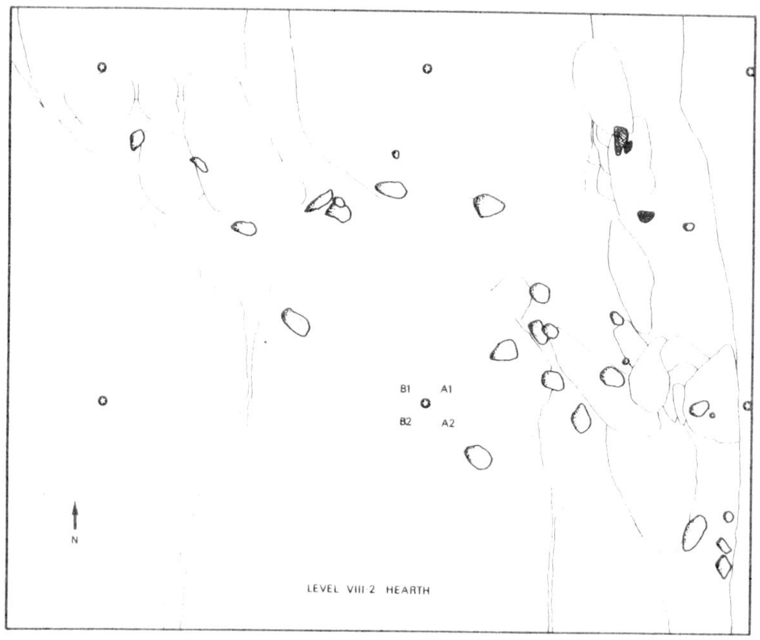

Figure 63. Plan of Cobble Concentration in Unit A1 (Hearth) for Le Flageolet I Level VIII-2.

heat (Robbins, personal communication). A third rich concentration is centered on unit B-1 (Figure 64). This accumulation is composed of small, fragmentary material and includes a few pieces of burnt limestone. As was the case for the A1 concentration, there is no evidence here for in situ burning. This deposit may represent hearth sweepings.

The three rich northern cobble accumulations are varied in form and content. They do not seem to reflect three separate fire features. Instead, they may be three subdivisions within a single "hearth complex". Interpretation of these clusters is difficult. However, the distribution of other artifactual materials clarifies the probable hearth position and the relationships among the three concentrations.

Density plots of cobble fragments less than 2 cm in maximum dimension are shown in Figure 65. Concentrations are apparant in units B0/B-1, which also contained two of the large cobble clusters. Unit A1, the richest large cobble accumulation, contains few fragments.

As was discussed for level VIII-1, hearth maintenance deposits generated by sweeping out features are expected to contain small cobble fragments and burnt sediments. It is unlikely, however, that they would comprise quantities of large, still useful cobbles. It must be remembered that all igneous rock must be brought to Le Flageolet from the river below. The concentration of large-sized pieces in unit A1, coupled with the linear arrangement of hearth-generated materials to the north, suggests that A1 may have been the locus of the hearth itself. Other deposits (i.e., those in B0/B-1) would then be sweepings from the feature. The sorting of these materials in the northern sector (with larger fragments in B0 near the hearth and smaller pieces in B-1 further away on the supporting rock) supports this hearth evacuation interpretation.

Small cobble fragments are also concentrated in southern units B6, B7, and C7. The area corresponds to the location of k-means cluster cobble centroids at both the two and seven cluster levels. Again crevices below the burnt rock surface contain these materials. The association in this space between large and small cobble fragments suggests that the rock surface was used during this occupation. However, it is the density of material that is impressive, and the relation is no less problematic than for level VIII-1.

Other artifact classes have density maps similar to that for small cobbles fragments. Bone debris, dechet, and debitage all have high frequencies in units B0 and B1, suggesting again that this accumulation was deposited by discard rather than as an in situ feature.

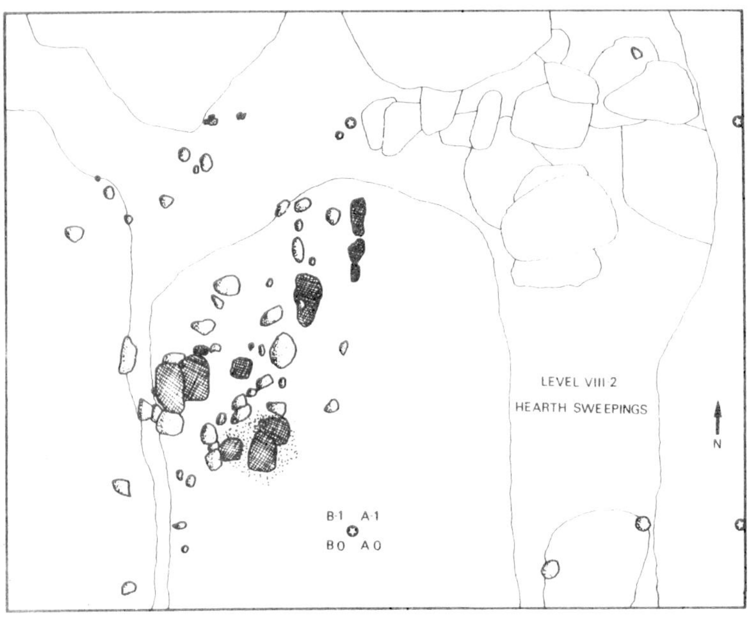

Figure 64. Plan of Cobble Concentration in Unit B-1 (Sweepings) for Le Flageolet I Level VIII-2.

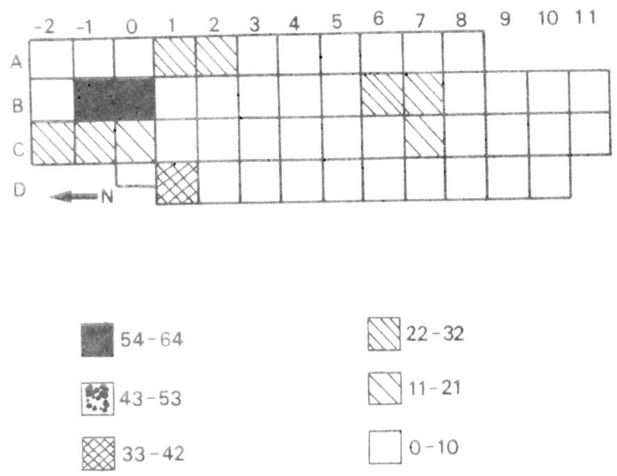

Figure 65. Density of Cobble Fragments (< 2cm) by Meter Square from Le Flageolet I Level VIII-2.

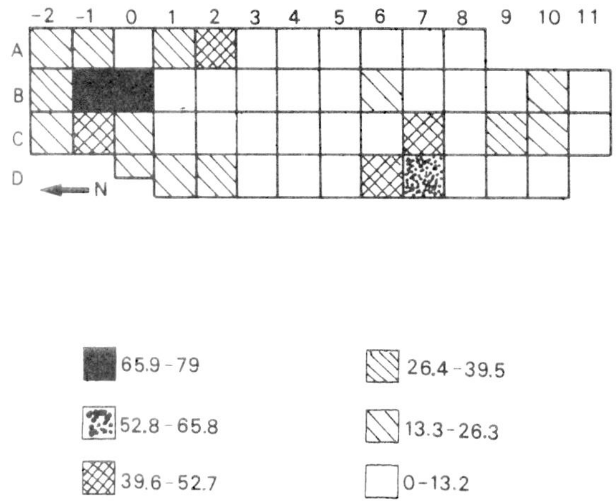

Figure 66. Density of Bone (> 2cm) by Meter Square from Le Flageolet I Level VIII-2.

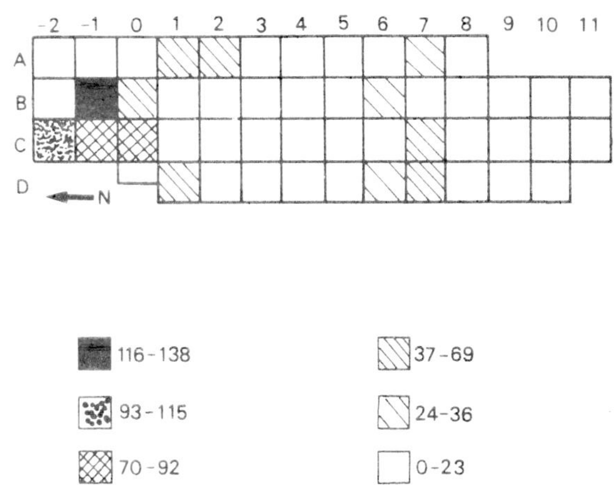

Figure 67. Density of Dechet (Lithic Materials < 2cm) by Meter Square from Le Flageolet I Level VIII-2.

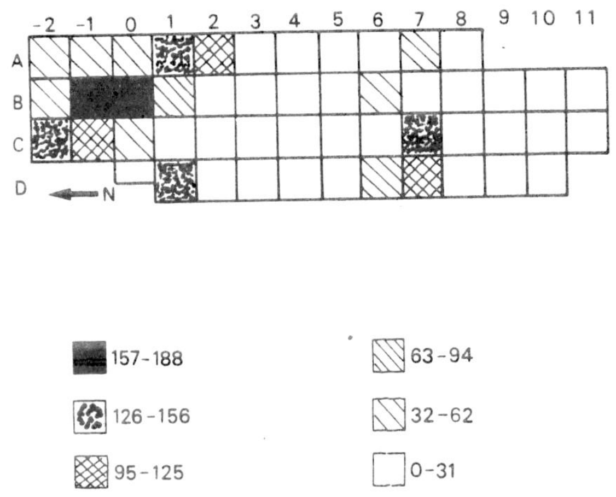

Figure 68. Density of Debitage (Lithic Materials > 2cm) by Meter Square from Le Flageolet I Level VIII-2.

Figure 66 shows bone counts by meter unit. Bones have their highest density in units B0 and B-1, but they are also concentrated along the west side of the southern burnt rock (units C7, D6/D7). A veritable bed of bone debris is present in the B0/B-1 area, but southern concentrations are spatially discrete.

Dechet also has its highest frequency in unit B-1 (Figure 67). Small lithic waste is abundant in unit C-2, and this concentration extends into adjacent units C-1 and C0. This is an area of high density for bone and small rock fragments. In the southern sector of the site, dechet is concentrated in units C7 and D7. A few pieces also occur in units A7 and B6. Again, the densest southern accumulations coincide among all grid provenienced artifacts.

Debitage, like the other level VIII-2 classes, has its richest concentration in units B0 and B-1 (Figure 68). Other rich northern sector units include C-2 (where it coincides with dechet but no other class), D1, and A1. The occurrence of larger lithic waste in the inferred hearth suggests again that reduction was performed at the feature and that waste was removed to discard spaces. Here, it would seem that some of the larger pieces were retained at the feature, perhaps for further modification. (Very few of these unmodified pieces are burnt and therefore were not knapped into the burning fireplace). In the southern sector, debitage is concentrated in units C7, D6, and D7. Again, this class overlaps with those just discussed.

Summary of VIII-2 Contextual Data

The overall picture of contextual elements deposited during the level VIII-2 occupation is somewhat more varied than was the case for VIII-1. Again, a single focal hearth feature is indicated by physical evidence, and this is located in the northern sector of the surface. A concentration of large cobbles in unit A1 may represent the actual hearth location. Two other cobble concentrations occur to the north and west of A1 along and on the surface of a large flat block that serves as a base for the feature complex. These accumulations contain cobbles, but rocks are smaller than in the A1 concentration. Small rock fragments, burnt limestone, bone debris, dechet, and debitage also compose the two rich northern accumulations. Particle sorting and the presence of burnt sediment and carbonized material in the B0 and B-1 concentrations suggest that these materials were swept from the hearth itself during maintenance operations.

A smaller, less rich area with a somewhat similar structure is present in the southern sector of the level VIII-2 surface below the burnt rock surface (B7). All grid provenienced data have high frequencies here, but these are consistently lower than in the north. There is no evident hearth feature in the southern sector. However, as was discussed above, these materials might be associated with a feature formerly present on the block surface above. If this was the case, then the content of the main southern accumulations should be similar to those around the northern hearth feature.

The presence of one, and possibly two, intensively-used hearths in level VIII-2 suggests a different kind of occupation from that reflected in level VIII-1. This difference may be due to several factors including population size, the length of occupation, the activities performed, or combinations of all of these. The spatial analysis of each of these surfaces, and comparisons between them, may allow the differences to be defined in more specific terms.

CHAPTER 7

LE FLAGEOLET I: LEVEL VIII-1

INTRODUCTION

A three stage spatial analysis was carried out for the Le Flageolet I level VIII-1 materials as for the Pincevent distributions (Chapter 4). First, point provenience data for ten different artifact classes were entered into separate k-means cluster analyses, and optimal cluster solutions examined for each class. Second, overlay maps were produced at three scales corresponding to optimal solutions indicated by the log(%SSE) plots. The maps were then inspected for patterning in order to define accumulation zones for association studies. Finally, after zones were identified, correlation studies and graphic comparisons were carried out.

Some General Expectations

Preliminary examinations of contextual elements in level VIII-1 (Chapter 6) can be used to generate expectations for the results of spatial analysis. A single hearth feature was identified, located in the northern sector of the site (units A1/B1). Based on the overlap of small rock fragments, dechet, debitage, and bone debris, a possible hearth dump was also indicated in the north (units C0/D0). The southern part of the site showed no firm evidence for a hearth feature, although there were cobbles present, and this area may represent a functionally different part of the occupation surface. Large block configurations should result in small, local concentrations of material confined to cracks between the blocks.

The level VIII-1 distributions should differ from the generalized accumulations at Pincevent. Near the focal hearth, artifact clusters are expected to form zones, and these should be relatively rich and contain a wide range of different classes. In the south, a more restricted set of artifact classes is anticipated. Zones in this sector should be related to specific activities occurring nearby. Given constraints to movement over the occupation surface, little or no mixing of northern hearth deposits with the southern sector materials is expected. The two sectors should retain their differences despite changing spatial refinement. Thus, change in homogeneity of deposits with spatial scale will be different from that observed at

Pincevent.

STAGE I ANALYSIS: PATTERN RECOGNITION

The ten artifact classes entered into *k*-means analysis for level VIII-1 comprise seven categories traditionally considered as "tools" and three of other lithic materials. As for the Pincevent data, each cluster analysis was performed to a fifteen cluster solution. Table 22 lists the classes and the number of each present in level VIII-1.

Log(%SSE) plots produced from the ten *k*-means runs show some variation in clustering sequences among the classes (Figure 69). In all cases, an optimal clustering solution is indicated when the distribution is broken into two clusters; as will be seen shortly, these correspond to the northern and southern sectors of the rockshelter. Beyond this level, different cluster solutions are optimal for different artifact classes. Excluding cobbles, which have already been discussed, most artifact classes have three optimal solutions. Several classes, including cores, truncated pieces, endscrapers, and notched pieces, show an optimal solution at between three and six clusters. The other classes have medium scale solutions with more clusters, i.e., eight or nine. High scale solutions are also indicated for all artifact catergories. All but one of these solutions represent "terminations" of the plots where the log(%SSE) decreases to zero. Table 23 lists the optimal solution levels for all classes indicated by the plots.

That the log(%SSE) should decrease to zero before the end of the run indicates a basic pattern in the data distributions for all classes. This pattern may relate to the number of small pockets within the rock fall that could contain material. A total of eleven to thirteen such pockets are suggested by the log(%SSE) terminations. Thus, it may be that the highest level patterns are contextually determined. One artifact class, debitage, does not terminate at zero before the fifteen cluster level is reached. This is the most abundant of all classes, and the likelihood is great of small clusters being defined within larger ones earlier in the clustering sequence. Nevertheless, the highest optimal solution for debitage is at twelve clusters, conforming to the highest solutions (number of pockets?) suggested by terminations in other categories. While the total number of pockets within rocks may be between eleven and thirteen, not all artifact classes are clustered in all pockets. High scale solutions range from eight clusters (for cores and notched pieces) to thirteen (for utilized pieces). This suggests, and will be confirmed below, that there are differences among classes in their distributions over the shelter surface at the highest spatial scale.

TABLE 22.

Frequencies of ten artifact categories analyzed for Le Flageolet I level VIII-1 assemblage.

ARTIFACT CLASS	N
Endscrapers	27
Burins	21
Aurignacian Pieces	26
Truncated Pieces	23
Notched and Denticulated Pieces	23
Miscellaneous Retouched Pieces	27
Utilized Pieces	47
Cores	24
Debitage	804
Cobbles and Fragments	21

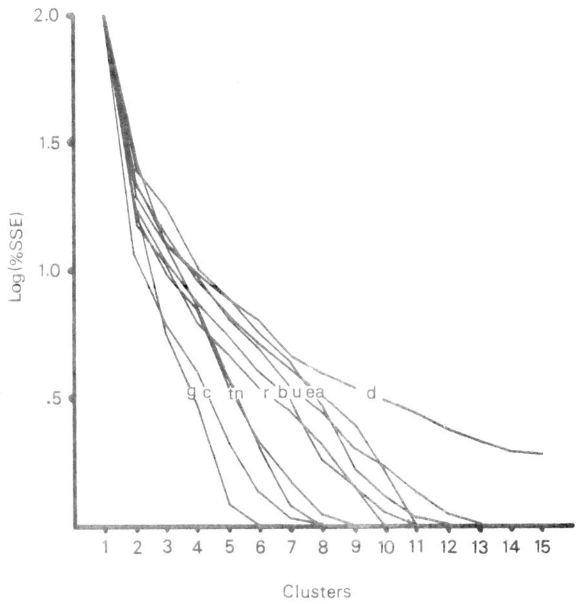

Figure 69. Plot of log(%SSE) Statistics for All Ten Artifact Classes in Le Flageolet Level VIII-1 as Defined by K-means Cluster Analysis to Fifteen Clusters.

TABLE 23.

Optimal clustering solution levels indicated by log(%SSE) plots produced by *k*-means analyses of Le Flageolet I level VIII-1 artifact distributions. A "t" indicates a terminal solution at zero before fifteen clusters are formed.

ARTIFACT CLASS	SOLUTION NUMBER 1	2	3	4	5
Cores	2	5	8t		
Cobbles	2	3	4	5	6t
Truncations	2	5	9t		
Burins	2	8	10t		
Endscrapers	2	4	9t		
Aurignacian Pieces	2	11t			
Retouched Pieces	2	9	11t		
Utilized Pieces	2	9	13t		
Notches and Denticulates	2	6	8t		
Debitage	2	8	12		

One artifact class, Aurignacian pieces, has only two optimal solutions indicated by the log(%SSE) plot. As for the other classes, a low level solution is at two clusters. The second solution is at the eleven cluster level where the log(%SSE) terminates. Aurignacian pieces may illustrate most precisely the basic contextual constraints on the VIII-1 materials. Northern and southern sectors, corresponding to a hearth area and a non-hearth area, are distinquished. At a fine scale, many small accumulations are defined within cracks between the large blocks. No other spatial patterning is exhibited by the Aurignacian classes.

To summarize the log(%SSE) plots, there appear to be two groups of artifact classes based on their clustering sequence. Both groups have optimal low level solutions at two clusters. One group has fewer clusters than the other at both medium and high clustering scales. Instead of describing each artifact class in turn, one from each of these two groups will illustrate the cluster formation sequence characteristic of all members. For the first group, truncated pieces are typical (with a two cluster, five cluster, and eight cluster solution) and will serve as an example. The second group will be exemplified by retouched pieces which have two, nine, and eleven cluster solutions.

Figure 70 shows the two cluster centroids and RMS circles defined by k-means analysis for truncated pieces at the lowest scale. One cluster occurs in the northern sector of the site near the A1/B1 hearth with its centroid in unit B0. The other cluster centroid is located on a large block (unit B9) in the southern sector. This large block shows evidence for burning and was discussed in the preceeding chapter. The northern cluster contains over 78% of all truncations present in level VIII-1. Only five are present in the southern cluster. For this artifact class, and for all classes except burins, the northern hearth is the center for the richest clusters.

A medium scale optimal solution is indicated for truncations at five clusters (Figure 71). One isolated piece is separated as an individual cluster, and this is located between the northern and southern sectors. Two clusters are present in each sector at this level. In the north, one centroid occurs in unit A0 (near the hearth) and the other in unit C0. In the southern sector, one cluster is located with its centroid in unit B8 not far from the two-cluster level centroid. A small cluster is located at the southern edge of the excavated area in unit C11. In content, the cluster centered in unit A0 contains eleven, or nearly 48%, of the truncated pieces. The cluster centered in C0 contains another six pieces (26% of the total). The same five pieces located in the south at the previous clustering level are divided between the two southern clusters at this scale.

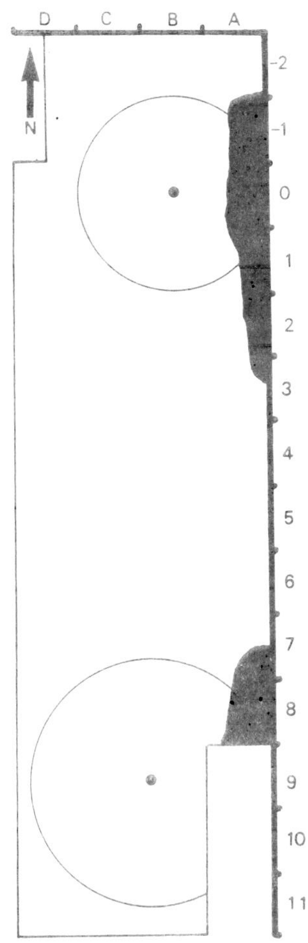

Figure 70. Map of Two Cluster K-means Solution for Le Flageolet I Level VIII-1 Truncation Distribution. Dots mark cluster centroids; circles are RMS radii for clusters.

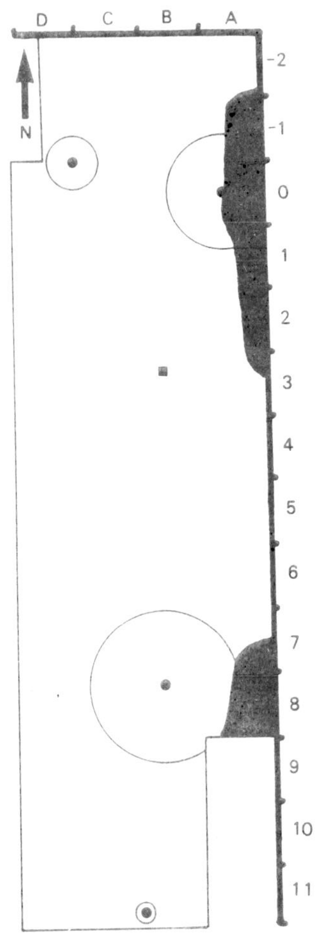

Figure 71. Map of Five Cluster K-means Solution for Le Flageolet I Level VIII-1 Truncation Distribution. Dots mark cluster centroids; circles are RMS radii for clusters.

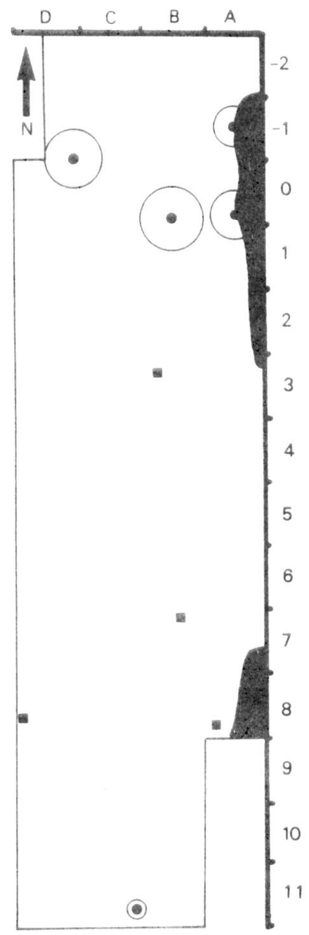

Figure 72. Map of Nine Cluster K-means Solution for Le Flageolet I Level VIII-1 Truncation Distribution. Dots mark cluster centroids; circles are RMS radii for clusters.

At the highest clustering scale, nine clusters are formed. Figure 72 illustrates the k-means map produced from the centroids and RMS measures provided by the run. First, note that many clusters at the highest scale are composed of isolated artifacts. Only in the northern sector, where abundance is greatest overall, are clusters comprising groups of objects indicated. Four clusters are present in the north. One of these is centered in unit A0 again, adjacent to the focal hearth. This is still a relatively rich cluster, containing six objects (26%). A second cluster is located at the line between units C0 and D0, and this also contains six truncations. A cluster of three pieces occurs to the north of the hearth area in unit A-1. A fourth northern sector cluster is located west of the hearth in unit B0 and represents a pair of truncated pieces.

A two-object cluster centered in unit C11 remains at this highest level, but the rest of the southern clusters at the previous level have been reduced to isolated pieces. The isolate in unit B3 between the northern and southern sectors is retained from the five cluster solution.

K-means analysis of truncated pieces indicates that a general pattern in the distribution is related to the A1/B1 hearth. Many truncations are centered right at the feature, while other concentrations occur just north and west towards the shelter exterior. A rich cluster is located between large blocks in units C0 and D0. As will be seen shortly, this crack is the center for tightly-packed centroids from many artifact classes. This characteristic supports the interprtation of this crack (as a hearth-area refuse dump) already proposed based on overlapping concentrations of small rock fragments, dechet, debitage, and bone.

Other artifact classes having the same general clustering sequence as truncations show patterns similar to that just described. Some of these other classes, cores for example, have southern sector clusters composed of sets of objects at higher scales. For the most part, northern sector clusters are the richest, are in proximity to the hearth, and overlap with small material concentrations. Southern sector clusters are made up of very few artifacts or a single object. It is important to note, however, that the number of clusters formed from level to level is roughly the same in the two sectors. This contrasts with the cluster formation process seen at Pincevent, where the richest areas of the site were divided first and into the most clusters as the runs proceeded. At Pincevent, this characteristic was linked to the abundance of materials in various parts of the site and to the homogeneity of accumulations regardless of their size. That the cluster formation process differs at Le Flageolet I (at least for level VIII-1) suggests that other kinds of relations may hold. This will be discussed in more detail later.

Miscellaneous retouched pieces provide an example of a second clustering pattern in level VIII-1. Figure 73 shows the map of clusters produced at the lowest k-means solution. As was the case for truncations, two clusters are present, one each in the northern and southern sectors. Nineteen pieces (over 70%) are assigned to the cluster centered on unit B0. The remaining eight miscellaneous retouched pieces are in the cluster centered on unit C8 west of the burnt block in the south. This is in contrast to truncations, which centered on the block itself.

Figure 74 illustrates cluster centroids and RMS radii at the nine cluster level. More clusters are formed at this medium scale than was the case for truncations, but these are again equally divided between the two sectors of the site surface. Five clusters are present in the north. Of these, one is centered in unit B0 just to the north of the hearth, a second is situated further north in unit A-1, and a third is located in the crack between blocks at the C0/D0 boundary. The first contains seven retouched pieces (over 25% of the total). The second contains only three pieces (11%). Two isolated pieces defined as clusters in units C-2 and A3 complete the divisions formed in the northern sector.

Four clusters are formed from the single low-scale cluster in the southern sector. One of these is an isolated piece in unit D11. The other clusters are centered in units D7, C7, and C9. The cluster in D7 abuts the edge of the excavated area. A total of seven retouched pieces are distributed more-or-less equally among the latter three clusters. In general, the richest clusters at the medium scale occur in the northern sector of the site, and equal numbers of clusters are formed in each zone from the initial two clusters. This formation pattern is identical to that noted for truncations, although more clusters are formed for retouched pieces at the medium scale.

The highest scale solution for retouched pieces identifies eleven clusters over the VIII-1 surface (Figure 75). Four multi-object clusters and three isolates are located in the northern sector. The same four southern clusters identified at the previous clustering scale are repeated here, including the isolated piece in D11. Isolated pieces in the northern sector are located in the A1 hearth, in unit A3 to the south of the feature, and in C-2. The single cluster located next to the hearth at the medium scale is here divided into two. One of these occurs in unit A0 and contains four retouched pieces (15% of the total). The other is situated to the west in B0 and is made up of three pieces (11%). Clusters are also centered near the shelter wall in A-1 (two pieces) and in the crack between rocks in D0 (seven objects or over

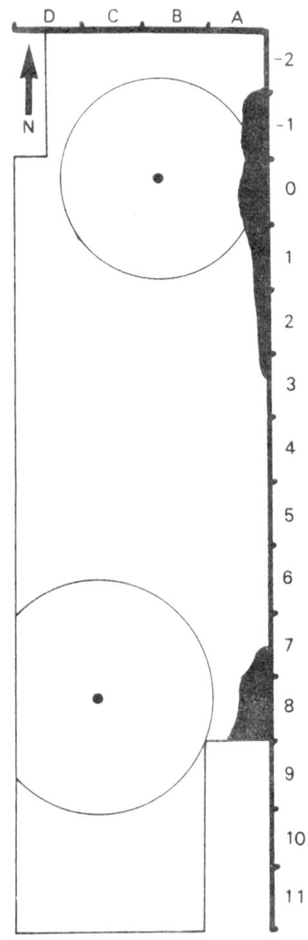

Figure 73. Map of Two Cluster _K_-means Solution for Le Flageolet I Level VIII-1 Retouched Pieces Distribution. Dots mark cluster centroids; circles are RMS radii for clusters.

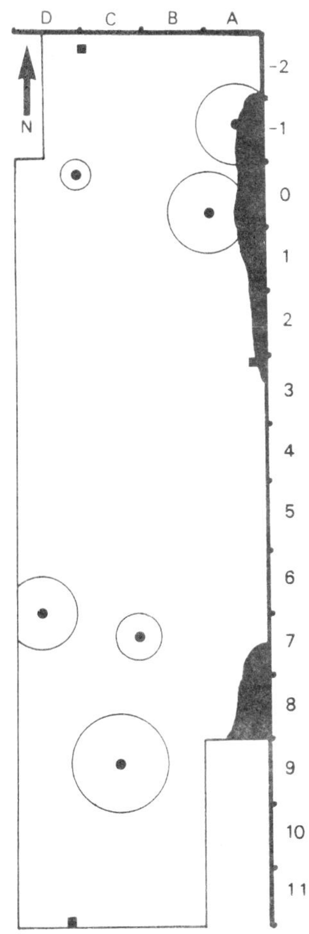

Figure 74. Map of Nine Cluster K-means Solution for Le Flageolet I Level VIII-1 Retouched Pieces Distribution. Dots mark cluster centroids; circles are RMS radii for clusters.

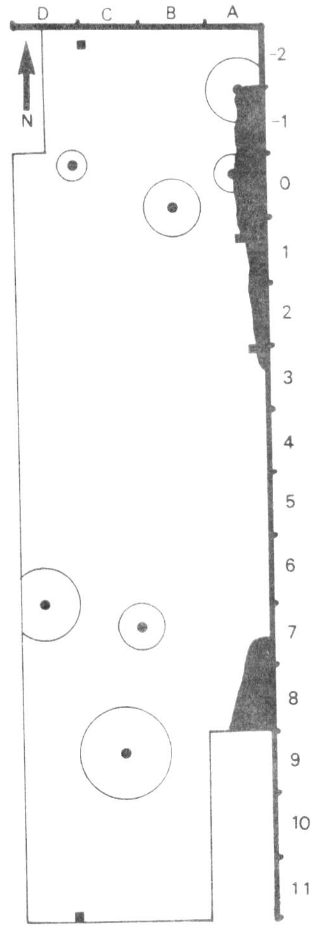

Figure 75. Map of Eleven Cluster K-means Solution for Le Flageolet I Level VIII-1 Retouched Pieces Distribution. Dots mark cluster centroids; circles are RMS radii for clusters.

one-quarter of the total).

One basic change occurs between the medium and high scale solutions for miscellaneous retouched pieces, relating to accumulations in the A1/B1 hearth area. The medium scale solution had a single, rich cluster centered on the feature itself. At the higher level, two clusters are formed at the hearth, but these are located next to the feature rather than in it. Only an isolated piece has its centroid in the hearth.

Some interesting characteristics of the level VIII-1 patterns are evident if the two sectors of the site are considered separately. The refinement of hearth-area (i.e., northern) distributions between medium and and high scales seems similar to that observed at Pincevent. The areas of greatest material density within this sector are divided into more clusters throughout the clustering sequence. This pattern is typical over the entire surface at Pincevent, however, while the southern sector of VIII-1 does not display this characteristic. That it occurs only on a local scale at Le Flageolet I, around the hearth, suggests a distinct global structure for the level. The difference is not surprising given the expectations for spatial patterning discussed at the beginning of this chapter. At both Pincevent and Le Flageolet I, dense accumulations of material near hearths seem to be those where overall material abundance contributes to cluster formation. A wide range of activities probably occurred at hearth features in both sites, resulting in the deposition of many artifact classes in large numbers. It is away from the hearths that a more heterogeneous deposition pattern seems to characterize the Le Flageolet level VIII-1 surface. This may only indicate that the distribution is well described after nine clusters are formed, and further clustering merely subdivides the hearth-area deposits. On the other hand, a basic structural aspect may be involved. This matter will be clarified as analysis proceeds.

The two general patterns of cluster formation, one progressing from two to around nine clusters with an intermediate solution of around five and another producing more clusters at the medium and high scales, together provide a unified picture of the VIII-1 distributions. Two sectors were defined based on unit-count data and shelter topography, and these are confirmed by the results of k-means analysis of ten artifact categories. As the sectors are subdivided according to individual artifact class distributions, each contains more-or-less equal numbers of clusters. This is true even though there is a discrepancy between the two general sectors in the absolute numbers of objects present. Nearly every artifact class has a termination solution at or near eleven clusters. Only debitage, the most abundant artifact class in the site, does not have a log(%SSE) plot that reaches zero before the run is complete. Still, this class has a final cluster solution at twelve clusters. The remarkable similarity among classes in both low and high scale clustering patterns suggests that these solutions may reflect contextual constraints in the overall distributions. Two-cluster solutions invariably locate clusters in each of the two flat spaces available for occupation, one in the north and one in the south. The rocky, uneven sector between the two flat ones is never the locus of a cluster centroid at the low scale. Highest level solutions are similar in number for all artifact classes and may reflect the absolute number of small pockets within the blocks that contain archaeological materials. While the basic pattern of pockets may be contextually defined, not all artifact classes are present in each pocket. This suggests that unconstrained patterning in the individual classes may be present at the highest clustering scale.

A last artifact class, Aurignacian pieces, will be used to illustrate the nature and locations of low and high scale patterns in a general fashion. As was discussed earlier, Aurignacian pieces are the sole class to have only two optimal clustering solutions. An examination of this artifact category may be useful as a visual means for appreciating the basic contextual situation in level VIII-1. For this reason, the maps produced from k-means analysis will be overlaid on the rock map presented earlier (as Figure 50).

The two-cluster solution for Aurignacian pieces illustrates the basic pattern (Figure 76). Twenty objects, or nearly 79% of the class, are located in a cluster centered near the hearth. The remaining six pieces are clustered in the south with their centroid in unit C7. The southern cluster lies in a crack just below the burned block.

Figure 77 presents the eleven cluster solution for Aurignacian pieces. This pattern is a rather complex scatter of clusters and isolates. In the southern sector, two clusters, each with two members, have centroids in units B7 and D8. Two isolated objects, one in D6 and one in C10, are also present in the south. In the northern sector, four clusters are defined. Three isolates (in C-2, A-1, and D2) are also indicated in the north. Two clusters are situated next to the hearth. One of these has its centroid in unit A1 and contains three pieces (11% of the total). The other, centered in B0, has eight pieces (over 30%). A third northern cluster is located in D0, and this has five members (19%). Finally, one other cluster is positioned between the two major sectors of the surface at this high scale. This cluster, comprising three artifacts, has its centroid in unit C3.

Overall, patterns illustrated by the k-means solutions for Aurignacian pieces clarify the view already suggested by other classes. Pockets of material are

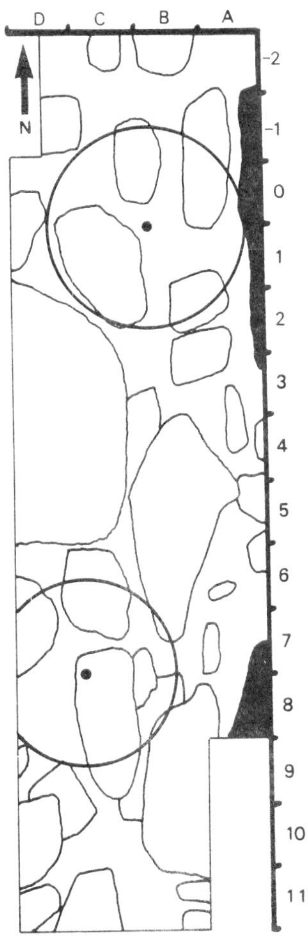

Figure 76. Map of Two Cluster K-means Solution for Le Flageolet I Level VIII-1 Aurignacian Pieces Distribution. Dots mark cluster centroids; circles are RMS radii for clusters.

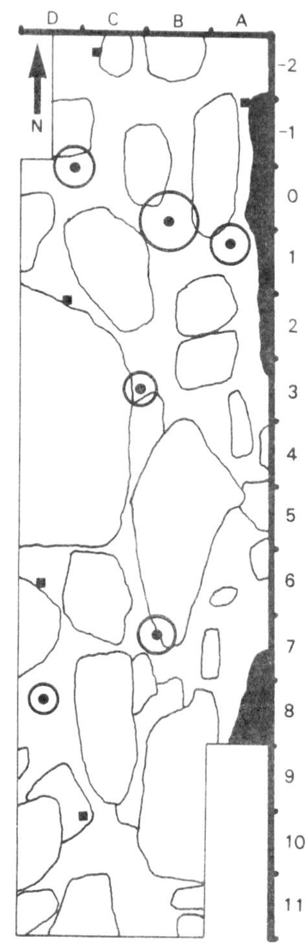

Figure 77. Map of Eleven Cluster K-means Solution for Le Flageolet I Level VIII-1 Aurignacian Pieces Distribution. Dots mark cluster centroids; circles are RMS radii for clusters.

indicated within two general sectors. In the north, pockets are associated with flat areas near the hearth and in a deep crack between rocks to the west of the hearth (units C0/D0). Very little material (mostly isolated artifacts) is present between the northern and southern sectors. Southern clusters are concentrated in a band of grid units, including B6, C7, C8, C9, D7, and D8. This band conforms to the base contour of the large block with a burnt surface. Cluster centroids around this block are located in the wider cracks between it and other nearby rocks. Within this general picture provided by Aurignacian pieces, other artifact classes have more complex patterns of deposition as evidenced by medium-scale cluster solutions. Medium patterns may reflect areas of deposition within the two general sectors but crosscutting the between-rock pockets.

The characteristics of individual artifact class distributions discussed here are suggestive of more general site patterning. To examine such patterning, the overlay technique employed in the Pincevent analyses will be applied. Three sets of *k*-means overlay maps are generated, corresponding to the three solution scales indicated for most artifact class distributions.

STAGE II ANALYSIS: PATTERN ASSOCIATION, ZONE DEFINITION

To compare the results of ten *k*-means analyses of individual artifact classes, three overlay maps are produced. Each map corresponds to one of the three solution scales indicated by individual runs. The lowest scale map overlays all of the two cluster solutions for level VIII-1 artifact categories (Figure 78). A medium scale map is produced by overlaying the medium level solutions for all the classes, and some variation in the number of clusters produced at this scale is present (Figure 79). The one artifact class showing only two optimal *k*-means solutions, Aurignacian pieces, must be treated somewhat differently than are the other classes. For Aurignacian pieces, the last, terminal solution will be mapped onto both the medium and high scale overlays. This procedure seems the most reasonable, since the several clusters present in a single zone can have their point frequencies summed to produce a zone frequency for the class. This combining of cluster point frequencies is employed for artifact classes having medium scale solutions when they have more than one cluster in a zone, so it is consistent with the entire analytic approach. Figure 80 shows the high scale overlay map for level VIII-1 and corresponds to the highest scale solutions (often terminal ones) for all artifact classes.

All two cluster solutions show a general coincidence across the level VIII-1 surface. A tightly packed group of centroids is present in the northern sector of the

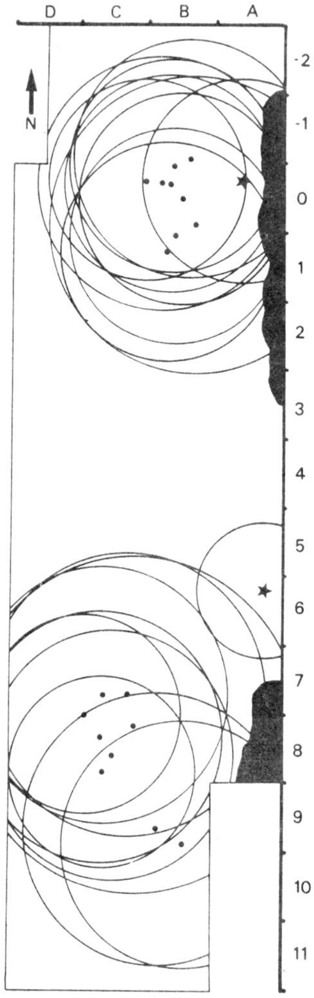

Figure 78. Low Scale Overlay Map Produced by K-means Analyses of Ten Artifact Class Distributions from Le Flageolet I Level VIII-1. Dots mark cluster centroids; boxes mark the position of isolated pieces; circles around centroids represent RMS radii for the clusters; cobble cluster centroids are marked with a star.

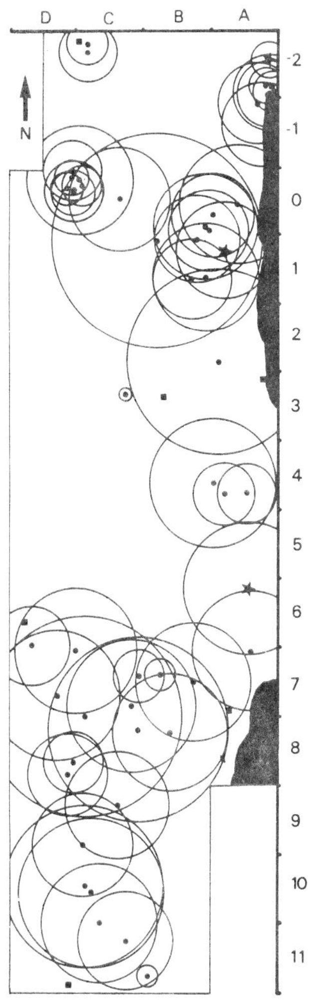

Figure 79. Medium Scale Overlay Map Produced by K-means Analyses of Ten Artifact Class Distributions from Le Flageolet I Level VIII-1. Dots mark cluster centroids; boxes mark the position of isolated pieces; circles around centroids represent RMS radii for the clusters; cobble cluster centroids are marked with a star.

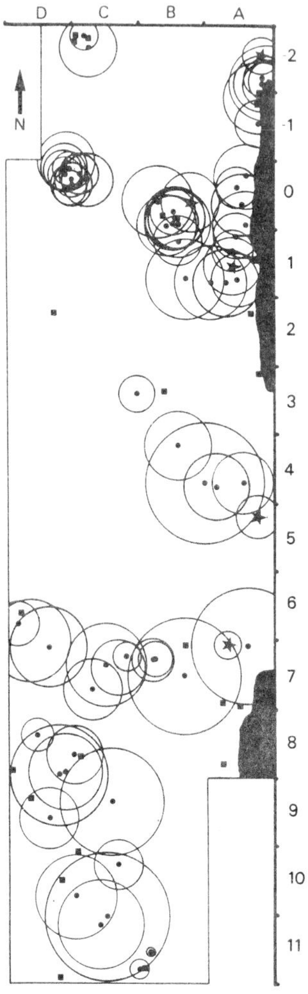

Figure 80. High Scale Overlay Map Produced by K-means Analyses of Ten Artifact Class Distributions from Le Flageolet I Level VIII-1. Dots mark cluster centroids; boxes mark the position of isolated pieces; circles around centroids represent RMS radii for the clusters; cobble cluster centroids are marked with a star.

shelter, and a more dispersed group is present in the south. In both sectors, the cluster closest to the shelter wall is composed of cobbles. All cluster centroids in the northern sector are located to the exterior side of the A1/B1 hearth (toward the opening of the rockshelter). The southern sector cobble cluster is positioned to the north and east of other artifact class clusters.

It is interesting to note that the northern clusters have an extremely compact centroid pattern. Cluster centroids for all ten artifact classes are located inside all RMS radii. This is not the case in the southern sector where two clusters, one each of truncated and notched pieces, lie to the south of the other centroids. The remaining southern sector clusters have a more compact grouping. Another important characteristic evident on this overlay map is a difference between northern and southern sector clusters in RMS size. Clusters in the north, near the hearth, have rather short radii indicating that they represent denser accumulations of material than do southern clusters. It has already been pointed out that northern sector clusters tend to be richer than southern ones, and this will be demonstrated later.

Two zones can be defined at the lowest clustering scale (Figure 81). These correspond to two areas of RMS overlap. The northern zone is centered on the inferred hearth in grid unit A1. The southern zone is composed of more dispersed clusters scattered over a wider surface; these are not associated with any evident hearth feature. This pattern conforms to that noted at Pincevent, where zones centered on focal hearths are compact and zones in peripheral areas are dispersed. It has already been suggested that the level VIII-1 surface contains a focal hearth in the northern sector and a more functionally-specific area in the south. The pattern of zones defined here supports such an interpretation of spatial organization in the shelter.

The medium scale overlay map generated from the ten k-means runs suggests a complex pattern of zones over the level VIII-1 surface. Three types of centroid dispersal are visible. In the southern part of the site, centroids are scattered in a rough arc around the large boulder centered in grid unit B7. No evident patterning can be seen in this southern cluster scatter, and a more-or-less continuous overlap of RMS radii covers the entire sector. Along the shelter wall between the north and south sectors, clusters are present but in reduced numbers. There is a single cobble cluster in this part of the surface, but few other artifact classes have nearby clusters at this scale. All clusters in the "middle sector" occur at the shelter wall. A very large block, at least thirty cubic meters in volume, dominates this part of the shelter and effectively closes the interior from the exterior. Middle sector clusters are confined to the

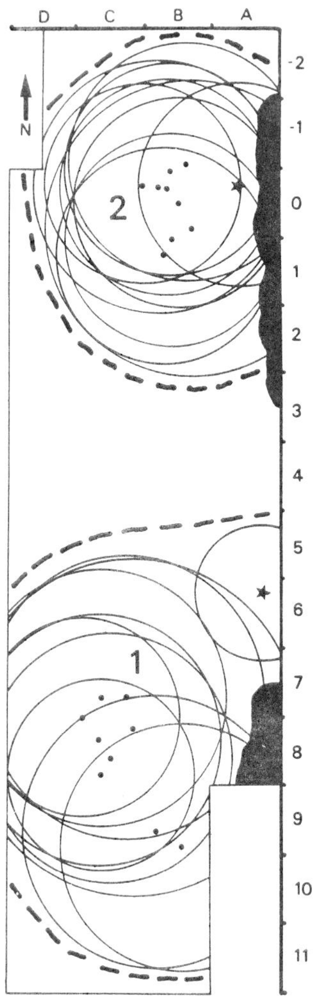

Figure 81. Two Zones Defined at Low Scale Solution for Ten Artifact Class Distributions from Le Flageolet I Level VIII-1. Zones are outlined with a dotted line.

space between this rock and the wall. The pattern of northern cluster locations is distinct at the medium scale. Near the hearth, several compact areas of centroid concentration are noted in marked contrast to the southern sector pattern. Many centroids occur just north of the hearth itself. Another group is located even further north against the shelter wall. A third compact centroid group is positioned to the west of the hearth, and all centroids are contained within all RMS radii. Two clusters and a single isolate are located to the north of this third group abutting the edge of the excavated area.

Eight zones are defined based on the medium scale overlay map (Figure 82). These differ in form and content, but they describe the pattern of centroid dispersal and RMS overlap discussed above. The southermost sector contains clusters whose RMS radii overlap widely over space, so this will be treated as a single zone at the medium scale. Near the shelter wall and to the north of zone 1, two clusters and an isolated piece make up a second zone defined because of its separation from other zones. Further north, a third zone is composed of three clusters with compact RMS overlap. A fourth zone is defined between zone 3 and the A1/B1 hearth; this zone 4 contains two clusters and two isolated pieces. Four more zones are located in the northern sector close to the hearth. Zone 5 is centered on the hearth itself. Zone 6 is a compact group of clusters against the shelter wall to the north of zone 5. The most compact zone is 7, which has concentrated centroids located in the C0/D0 rock crack. Finally, two clusters and one isolate located at the northwestern edge of the excavated area are collected into zone 8.

The distribution of medium scale zones over the level VIII-1 surface refines some of the observations made at the lower solution scale. New information is also obtained. Different patterns for northern and southern sectors, suggested at the low scale, are confirmed. Clusters in the southern sector are more dispersed than in the north--so much so that no clear zones can be defined within the sector. Cluster centroids seem to have a linear pattern in the south conforming to the base of a large boulder. It may be that patterning in the southern sector will be detected at finer scales, but none is evident here. Against the shelter wall, a middle sector contains a sparse scatter of clusters and isolates. These may represent small pockets of material lodged in cracks within the chaotic rockfall in the site's central portion. Northern sector zones are more compact and better-defined than those in the middle and south. One rich zone is associated with the hearth in A1/B1 but is located on only one side of the feature. Three other zones, some confined to cracks between rocks, compose a fan-shaped pattern around the hearth.

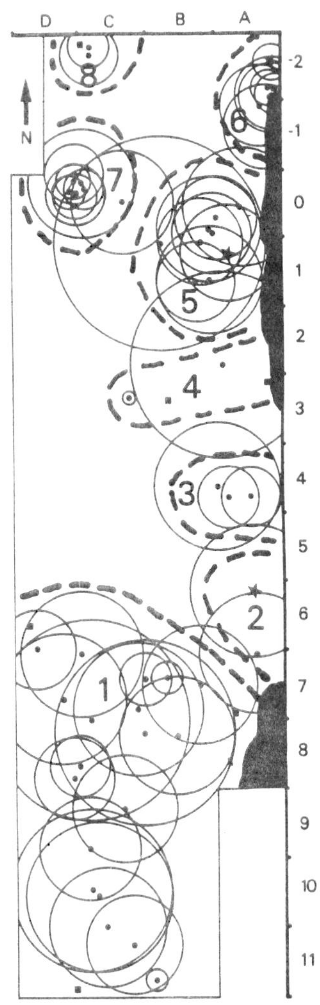

Figure 82. Eight Zones Defined at Medium Scale Solution for Ten Artifact Class Distributions from Le Flageolet I Level VIII-1. Zones are outlined with a dotted line.

Zones for the high scale overlay map are defined on Figure 83. This map clarifies the patterns already noted at lower scales, especially in the southern sector. Zones in the middle and northern sectors are also refined, although differences between this and the medium solution scale are not as great as in the south. The implications of this refinement difference will be discussed below.

At the highest clustering scale, five zones can be distinguished in the northern part of the level VIII-1 surface. Only one of the medium scale zones is changed in this sector, and that is the one associated directly with the hearth. Designated zone 5 at the medium scale, the hearth area now has two separate zones. One of these is a linear configuration of centroids passing from grid unit A1 along the shelter wall to unit A0. The form of this zone is constrained by a crack between rocks that dominates units B1 and B0 and extends into units A1 and A0. A second hearth area zone is centered in grid unit B0. This zone is more compact than that in A1 with all centroids tightly concentrated within RMS radii. Other zones identified at the previous clustering scale are retained in the north sector. There is a zone centered on grid unit C0, another in A-2, and a third in unit C-2.

In the middle sector, two zones are identified at this scale. One conforms to zone 3 at the medium solution level, and the other comprises a cluster and an isolate with centroids located in a crack between two large blocks.

In the southern sector, cluster centroids again are concentrated along the base of a large boulder and have a "Y" -shaped pattern. At this finest clustering scale, five zones can be discerned within this pattern. At the extreme southern end of the excavated area, a small concentration of two clusters and an isolate is designated zone 1. Moving north, two distinct zones consisting of various cluster and isolate combinations are defined as zones 2 and 3. At the northern end of the large burnt rock, another linear scatter of clusters is collected as zone 4. Although there seem to be more clusters concentrated at one corner of this block, a few other clusters and isolates closer to the shelter wall are included in the zone because RMS overlap is continuous across the band. Just to the west of zone 4, two clusters and an isolate are designated zone 5; these have common RMS overlap that extends only slightly into zone 4 without encompassing any of the latter's cluster centroids.

The overall pattern of cluster centroids and RMS radii at the highest clustering scale provides a detailed picture of zone distributions over the level VIII-1 surface. The northern sector is characterized by dense and compact material accumulations positioned around the only certain hearth feature in the occupation. Some zones are located along the shelter wall to the north of the

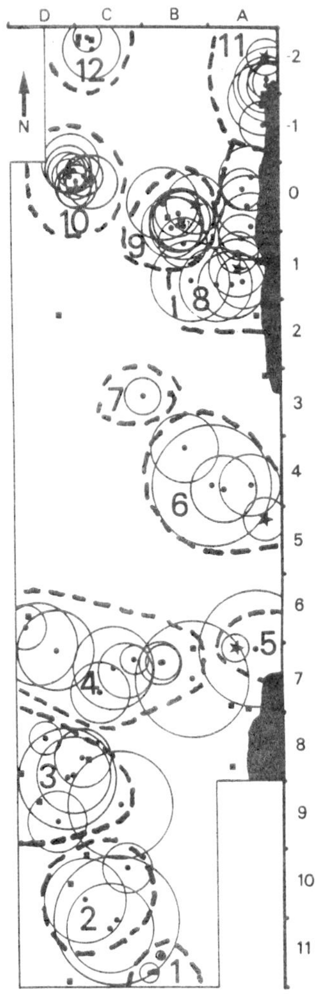

Figure 83. Twelve Zones Defined at High Scale Solution for Ten Artifact Class Distributions from Le Flageolet I Level VIII-1. Zones are outlined with a dotted line.

hearth. Others occupy cracks between rocks to the north and northwest of the feature. The overall pattern is similar to the focal hearth deposits at Pincevent, with dense material accumulations forming an arc on one side of the feature. However, the topographic situation at Le Flageolet may have constrained hearth area deposition so that dumps were located in "holes" between rocks rather than in an unrestricted fan.

In the southern sector of the site, discrete cluster zones only become apparant at the highest clustering scale. The basic pattern of deposits is linear and follows the contours of a single large block. That this pattern can only be subdivided at the highest scale of resolution suggests that southern sector accumulations are (1) quite small and (2) quite different that those defined around the hearth. Given the nature of southern sector distributions, it seems plausible to suggest the use of block surfaces as activity spaces in this sector (in particular the wide, flat surface of the B7-B10 boulder). If this was so, then the "activity areas" were suspended above the resulting material desposits. That discrete zones of accumulation can be defined around the block is encouraging since these pockets may reflect the activites that occurred above.

In the middle sector of the surface, refuse accumulations are diffuse and sparse. Two zones seem to represent isolated pockets of material lodged within the rock jumble dominating this part of the site. These zones may reflect objects dropped or lost as site occupants moved over the central sector.

While the spatial patterns revealed by pattern recognition techniques and zone definition are suggestive, the meaning of apparant organizational differences (between sectors and among zones) cannot be addressed until contents have been examined. The zones defined here will serve as the units for content analysis.

STAGE III ANALYSIS: PATTERN ASSOCIATION, CONTENT ANALYSIS

Assessing Global Associations among Artifact Classes

As for the Pincevent data, Spearman's rho statistic for rank order correlation is computed among the ten artifact classes analyzed for level VIII-1. Tables 24, 25, and 26 present the raw frequencies and within-class rank values assigned to all zones at the low, medium, and high scale solutions for level VIII-1, respectively.

Table 27 shows the rank order correlation matrix for the level VIII-1 low scale solution. The northern zone (zone 2) contains all of the highest ranking clusters for all artifact classes except burins. That burins should be

TABLE 24.

Raw counts and ranks for ten artifact classes from Le Flageolet level VIII-1 based on zones defined at lowest clustering solution scale. Rank values are in parentheses.

MATERIAL CLASS	CLUSTER ZONE 1	CLUSTER ZONE 2
Cores	10 (2)	14 (1)
Cobbles	6 (2)	15 (1)
Truncations	5 (2)	18 (1)
Burins	11 (1)	10 (2)
Endscrapers	7 (2)	20 (1)
Aurignacian Pieces	6 (2)	20 (1)
Retouched Pieces	8 (2)	19 (1)
Utilized Pieces	17 (2)	30 (1)
Notched Pieces	4 (2)	19 (1)
Debitage	185 (2)	619 (1)

TABLE 25.

Raw counts and ranks for ten artifact classes from Le Flageolet I level VIII-1 based on zones defined at the medium clustering solution scale.

MATERIAL CLASS	CLUSTER ZONE 1	2	3	4	5	6	7	8
				COUNTS				
Cores	10	0	0	0	3	6	5	0
Cobbles	0	6	0	0	10	5	0	0
Truncations	5	0	0	1	11	0	6	0
Burins	11	0	0	2	5	1	2	0
Endscrapers	6	0	0	5	0	5	11	0
Aurignacian	6	0	0	4	11	1	3	1
Retouched	8	0	0	1	7	3	7	1
Utilized	12	3	2	0	12	6	8	4
Notched	3	1	3	0	7	4	5	0
Debitage	161	0	25	0	211	75	294	36
				RANKS				
Cores	1	7.5	7.5	7.5	4	2	3	7.5
Cobbles	7.5	2	7.5	7.5	1	3	7.5	7.5
Truncations	3	7.5	7.5	4	1	7.5	2	7.5
Burins	1	7.5	7.5	3.5	2	5	3.5	7.5
Endscrapers	2	7.5	7.5	3.5	7.5	3.5	1	7.5
Aurignacian	2	7.5	7.5	3	1	5.5	4	5.5
Retouched	1	7.5	7.5	5.5	2.5	4	2.5	5.5
Utilized	1.5	6	7	7.5	1.5	4	3	5
Notched	4.5	6	4.5	7.5	1	3	2	7.5
Debitage	3	7.5	6	7.5	2	4	1	5

TABLE 26.

Raw counts and ranks for ten artifact classes from Le Flageolet I level VIII-1 based on zones defined at the highest clustering solution scale.

MATERIAL CLASS	CLUSTER ZONE 1	2	3	4	5	6
			COUNTS			
Cores	0	2	3	4	0	0
Cobbles	0	0	0	3	0	3
Truncations	2	0	1	1	0	0
Burins	1	2	3	4	1	0
Endscrapers	0	1	1	4	0	2
Aurignacian	0	1	2	2	1	0
Retouched	0	0	2	3	2	0
Utilized	3	1	4	5	2	2
Notched	0	2	1	0	0	3
Debitage	0	20	37	61	43	24
			RANKS			
Cores	11.5	5.5	4	3	11.5	11.5
Cobbles	11.5	11.5	11.5	4.5	11.5	4.5
Truncations	4.5	11.5	5.5	5.5	11.5	11.5
Burins	4.5	3.5	2.5	1	4.5	11.5
Endscrapers	11.5	6.5	6.5	3	11.5	5
Aurignacian	11.5	5.5	4.5	4.5	5.5	11.5
Retouched	11.5	11.5	4.5	3.5	4.5	11.5
Utilized	7	10	5.5	4	8.5	8.5
Notched	11.5	5	6	11.5	11.5	4
Debitage	11.5	10	7	5	6	9

TABLE 26 (Continued).

MATERIAL CLASS	CLUSTER ZONE 7	8	9	10	11	12
			COUNTS			
Cores	0	0	2	5	6	0
Cobbles	0	6	4	0	5	0
Truncations	1	6	2	6	3	0
Burins	2	3	2	2	1	0
Endscrapers	0	3	10	0	5	1
Aurignacian	3	3	8	3	1	1
Retouched	0	5	3	7	2	1
Utilized	0	8	6	8	4	4
Notched	0	7	0	5	4	0
Debitage	0	86	134	294	67	36
			RANKS			
Cores	11.5	11.5	5.5	2	1	11.5
Cobbles	11.5	1	3	11.5	2	11.5
Truncations	5.5	1.5	4.5	1.5	3	11.5
Burins	3.5	2.5	3.5	3.5	4.5	11.5
Endscrapers	11.5	4	1	11.5	2	6.5
Aurignacian	3.5	3.5	1	3.5	5.5	5.5
Retouched	11.5	2	3.5	1	4.5	5.5
Utilized	11.5	1.5	3	1.5	5.5	5.5
Notched	11.5	1	11.5	2	3	11.5
Debitage	11.5	3	2	1	4	8

TABLE 27.

Matrix of Spearman's rank order correlation statistics (rho) for all Le Flageolet I level VIII-1 artifact classes when two cluster zones are considered. Artifact classes are numbered as follows: (1) cores, (2) cobbles, (3) truncations, (4) burins, (5) endscrapers, (6) Aurignacian pieces, (7) miscellaneous retouched pieces, (8) utilized pieces, (9) notched pieces, and (10) debitage. All correlations of 1.0 have ($p < .01$).

	1	2	3	4	5	6	7	8	9	10
1	/									
2	1.0	/								
3	1.0	1.0	/							
4	-1.0	-1.0	-1.0	/						
5	1.0	1.0	1.0	-1.0	/					
6	1.0	1.0	1.0	-1.0	1.0	/				
7	1.0	1.0	1.0	-1.0	1.0	1.0	/			
8	1.0	1.0	1.0	-1.0	1.0	1.0	1.0	/		
9	1.0	1.0	1.0	-1.0	1.0	1.0	1.0	1.0	/	
10	1.0	1.0	1.0	-1.0	1.0	1.0	1.0	1.0	1.0	/

separate from the other classes at this scale, having their greatest frequency in the southern sector, seems to suggest that this tool was often used away from hearth contexts. Referring back to the raw counts for this pattern (Table 24), the difference in number of burins between the two zones is not great; however, discrepancies for other artifact classes are much wider than for burins.

Table 28 shows rank order correlations at the medium scale for level VIII-1. Here, only four pairs of classes have values that are significant at the $(p < .05)$ level. Ten classes are intercorrelated if values are considered significant at $(p < .1)$. The only classes that have no significant correlations regardless of stipulated $\underline{p}$ values are cobbles and endscrapers. Based on the lower significance value, two groups of artifact classes can be defined. The first group comprises cores, miscellaneous retouched pieces, utilized pieces, and burins. This group has highest ranking clusters in the southern sector at the medium scale. Classes associated in the northern sector are truncations, cobbles, Aurignacian pieces, utilized pieces, notched pieces, miscellaneous retouched pieces, and debitage. As can be seen, utilized pieces and miscellaneous retouched pieces have weak associations with other classes in both sectors of level VIII-1.

When the stipulated significance level of $(p < .05)$ is considered, two groups can again be defined. Classes with their highest cluster ranks in the southern sector are burins, Aurignacian pieces, and miscellaneous retouched pieces. Classes associated primarily in the northern sector include miscellaneous retouched pieces, utilized pieces, and debitage. Here, miscellaneous retouched pieces have significant relations with different classes in each sector of the level VIII-1 surface.

Of the artifact classes having their highest ranking clusters in the northern sector, all but endscrapers and debitage are most common in zone 5. The two exceptions are best represented in zone 7. In the southern sector, rank patterning is not quite as clear. For example, burins and Aurignacian pieces have their highest ranks in different zones; still, they are correlated at $(p < .05)$. This relation indicates that the distributions of these two classes are similar over the whole surface. Intercorrelation between the two sector-based artifact groups suggests a complicated set of relations. However, links are usually through a single artifact class (miscellaneous retouched pieces at the $(p < .05)$ level). The most important characteristic of level VIII-1 correlation is the reduced number of relationships defined after one round of spatial refinement. At the low scale, all artifact classes but one were perfectly intercorrelated. Here at the medium scale, there are only four correlations that are significant at the stipulated level. This is in contrast to the situation at Pincevent, where similar numbers of relations were detected

TABLE 28.

Matrix of Spearman's rank order correlation statistics (*rho*) for all Le Flageolet I level VIII-1 artifact classes when eight cluster zones are considered. Artifact classes are numbered as follows: (1) cores, (2) cobbles, (3) truncations, (4) burins, (5) endscrapers, (6) Aurignacian pieces, (7) miscellaneous retouched pieces, (8) utilized pieces, (9) notched pieces, and (10) debitage. *Rho* values with ($p < .05$) are starred.

	1	2	3	4	5	6	7	8	9
1	/								
2	-.22	/							
3	.33	-.35	/						
4	.65	-.30	.84	/					
5	.57	-.07	.28	.53	/				
6	.43	-.23	.83	.92*	.29	/			
7	.82	-.29	.70	.93*	.57	.83	/		
8	.77	.05	.56	.68	.22	.68	.90*	/	
9	.59	.20	.45	.44	.08	.38	.58	.72	/
10	.71	-.01	.58	.57	.33	.85	.85	.88*	.81

throughout the refinement sequence.

Table 29 shows the correlation matrix obtained when twelve zones are considered. Only three correlation values are significant at this highest solution scale ($p < .05$). These form a single, inter-related set of classes. Composing the group are utilized pieces, miscellaneous retouched pieces, and debitage. Two of these classes (miscellaneous retouched pieces and debitage) have their highest frequencies in zone 8 centered on the hearth. Utilized pieces have their highest ranks in zones 8 and 10. Even artifact classes having correlation values that are significant only at ($p < .1$), e.g., cobbles and endscrapers, truncations and utilized pieces, have their highest ranks in these two zones.

Overall, the number of significant correlations decreases between each scale of spatial refinement. A general pattern of intercorrelation is present at the lowest level (two zones). At the medium scale, with the distributions divided into eight zones, fewer correlations are detected, and these occur within groups of classes concentrated in the north and the south sectors. At the highest clustering scale, very few relationships are significant, and these all occur among classes concentrated in the northern hearth area. This global association pattern suggests that southern sector zones (once they have been defined at the high solution scale), contain restricted sets of artifact classes.

The basic division in level VIII-1 between the northern hearth area and the southern sector extends beyond the abundance of materials to the kinds of artifacts occurring in each part of the surface. Around the hearth, deposits seem to correspond to those noted at Pincevent in terms of spatial organization and density of material accumulations. However, a limited set of classes is associated in the northern zones at Le Flageolet. More diffuse spatial patterns and organization characterize the level VIII-1 southern sector. This is in marked contrast to Pincevent, where abundance-related interclass associations are present over the entire site surface. In short, heterogeneity in accumulation content is indicated at the finer spatial scales in level VIII-1, and this probably relates to the specific nature of the southern sector.

Figure 84 illustrates the homogeneity of zones plotted against solution scale. This graph is quite different than that produced for the Pincevent data (see Figure 31 in Chapter 4) where a more-or-less flat line was evident across refinement scales. The plot for level VIII-1 begins with a very homogeneous profile when only two zones are defined. As the scale becomes finer (as more zones and smaller zones are distinguished), accumulations become more heterogeneous. This trend

TABLE 29.

Matrix of Spearman's rank order correlation statistics (*rho*) for all Le Flageolet I level VIII-1 artifact classes when twelve cluster zones are considered. Artifact classes are numbered as follows: (1) cores, (2) cobbles, (3) truncations, (4) burins, (5) endscrapers, (6) Aurignacian pieces, (7) miscellaneous retouched pieces, (8) utilized pieces, (9) notched pieces, and (10) debitage. *Rho* values with ($p < .05$) are starred.

	1	2	3	4	5	6	7	8	9
1	/								
2	.22	/							
3	.17	.04	/						
4	.05	-.54	.45	/					
5	.15	.68	-.06	-.18	/				
6	.43	-.57	.02	.30	.05	/			
7	.20	-.06	.38	.20	.13	.26	/		
8	.22	.19	.62	.31	.37	.15	.88*	/	
9	.01	-.00	.06	-.45	-.09	-.63	.11	.14	/
10	.38	.28	.48	.23	.37	.35	.90*	.83*	.18

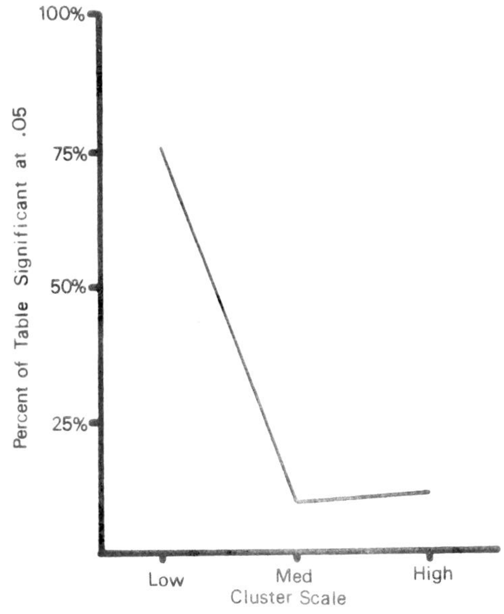

Figure 84. Plot of Accumulation Content Homogeneity with Scale of Spatial Refinement for Le Flageolet I Level VIII-1.

conforms to the pattern predicted when discrete activity areas are preserved in the spatial record. This, in turn, indicates that site formation processes during the level VIII-1 occupation differed from the organized dumping inferred for Pincevent. In the hearth area, dumps do seem to exist in VIII-1. These are similar to those around the Pincevent focal hearths in that they contain large quantities of material and are generally positioned in an arc to one side of the central feature. However, only a few artifact classes are associated in the hearth area zones at Le Flageolet, and these may reflect a limited number of tools used in hearth contexts. Other refuse deposits not associated with the A1 hearth are heterogeneous, and refuse disposal away from the feature context seems to have been more direct, i.e., specific activity sets retained their integrity. This may have been due to difficulties in moving refuse over the site surface created by the large and uneven rock surfaces. If specific activities were carried out on the surfaces of the southern sector blocks, then the results of the activities may have been dumped into a nearby crack rather than transported to centralized refuse piles. Thus, variation between northern and southern zones may reflect local content differences between hearth-centered group maintenance contexts and functionally specific areas. To examine variation in zone contents on a local level, zones must be examined individually and compared to determine if such patterning exists.

Assessing Local Associations among Artifact Classes

Table 30 shows the class maximum standardized matrix for the level VIII-1 low scale solution. As was indicated during correlation analysis, all artifact classes except burins achieve their highest values in northern zone 2. Burins have their highest class maximum value in zone 1. Of the other nine artifact classes, only cores have even a relatively high value in zone 1. This table indicates that most artifacts are concentrated in the north near the hearth and that the relative abundance of most classes is related to location at the lowest solution scale. Such a relationship was taken as indicative of systematic refuse disposal at Pincevent, and this interpretation may be relevant here insofar as disposal was confined to broadly conceived sectors of the surface. However, as we shall see, refuse mixture <u>between</u> sectors does not appear to have been frequent.

Figure 85 shows the circular graphs produced from Table 30. The concentration of all artifact classes in northern zone 2 is clear. Even the one class with a value of 100 in zone 1, burins, has an almost equal value in zone 2. Zone 1, on the other hand, is quite variable in content. High values are present for burins and cores, and the radii corresponding to cobbles and utilized pieces

TABLE 30.

Matrix of class maximum standardization values for Le Flageolet I level VIII-1 artifact classes when two cluster zones are considered. All values are rounded to the nearest integer.

MATERIAL CLASS	CLUSTER ZONE 1	2
Cores	71	100
Cobbles	40	100
Truncations	28	100
Burins	100	91
Endscrapers	35	100
Aurignacian	30	100
Retouched	42	100
Utilized	57	100
Notched	21	100
Debitage	30	100

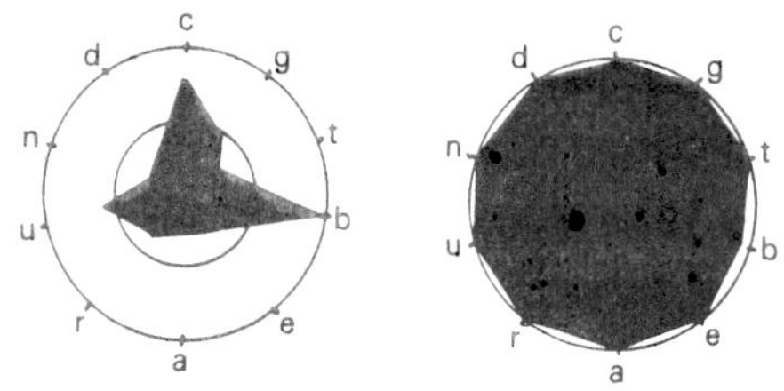

Figure 85. Circular Graphs for Class Maximum Standardization of Zone Contents when Two Zones Are Considered for Level VIII-1. Radii correspond to artifact classes as follows: (c) cores, (g) cobbles, (t) truncations, (b) burins, (e) endscrapers, (a) Aurignacian pieces, (r) miscellaneous retouched pieces, (u) utilized pieces, (n) notched pieces, (d) debitage. Circumference of circle represents class maximum value of 100.

also have peaks. At this lowest scale the two zones seem to represent two local content varieties. Zone 2, near the hearth, has a profile similar to those produced by hearth-centered zones at Pincevent. Zone 1 is different from zone 2 and resembles the small, peripheral accumulations at Pincevent. This sructural difference was noted as correlation tables were examined, but the nature of the variation between low scale zones could not then be defined. Now we can see that three artifact classes are evenly distributed between the two zones--cores, burins, and utilized pieces. Other classes are concentrated in the hearth centered northern zone.

To characterize zones independently of material abundance, zone maximum standardization is performed (Table 31). Debitage is excluded from these within-zone calculations because it so dominates both zones. The two low scale zones are similar for several classes (cores, cobbles, endscrapers, miscellaneous retouched pieces, and utilized pieces). Differences between zones also exist. Zone 1 is characterized by a high burin percentage, and zone 2 has high values for truncations, Aurignacian pieces, and notched pieces. Graphic depictions highlight these differences (Figure 86). The graph for zone 1 shows a peak at the burin position; that for zone 2 has a peak for truncations. Other aspects of these graphs are similar.

Two axes of variation describe zone contents at the low scale. First, artifact density explains some of the differences between zones 1 and 2, although several classes--most notably burins, cobbles, and utilized pieces--do not seem to be affected by abundance in the same way as other classes. A second dimension concerns zone composition. In content, the two zones are remarkably similar. Burins concentrate in the southern zone and truncated pieces are strongly represented near the northern focal hearth. Other artifact classes seem to follow the same distribution pattern in both zones, indicating equivalent units in terms of general composition.

When eight cluster zones are considered, standardized values yield a different pattern of artifact distribution. Table 32 shows the class maximum values for the level VIII-1 medium scale solution. Zone 1, representing the entire southern sector, has maximum values for four artifact classes: cores, burins, retouched, and utilized pieces. Two of these, retouched and utilized pieces, also have high class percentages in other zones. The other two classes have middle values in other zones. Zone 5 contains high values for a number of classes; this is the zone centered on the A0/A1 hearth feature. Here, cobbles, truncations, Aurignacian, utilized, and notched pieces have their highest within-class percentages. Debitage and miscellaneous retouched pieces also have relatively high values in zone 5. Zone 7, centered on units C0/D0, shows

TABLE 31.

Matrix of zone maximum standardization values for Le Flageolet I level VIII-1 artifact classes when two cluster zones are considered. All values are rounded to the nearest integer. Note that debitage has been excluded from calculations.

MATERIAL CLASS	CLUSTER ZONE 1	CLUSTER ZONE 2
Cores	59	47
Cobbles	35	50
Truncations	29	60
Burins	65	33
Endscrapers	41	67
Aurignacian	35	67
Retouched	48	63
Utilized	100	100
Notched	24	63

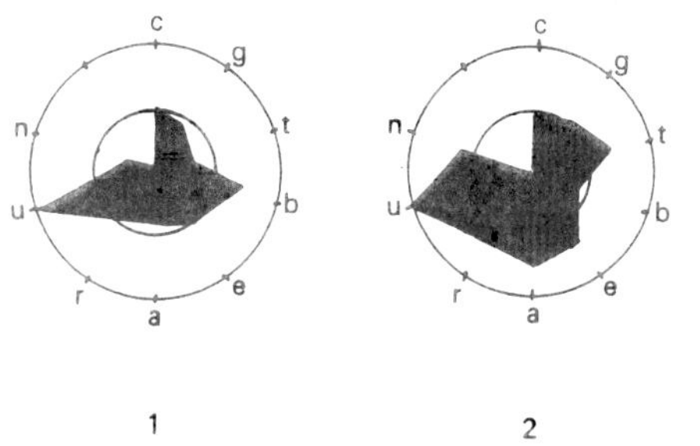

Figure 86. Circular Graphs for Zone Maximum Standardization of Zone Contents when Two Zones Are Considered for Level VIII-1. Radii correspond to artifact classes as follows: (c) cores, (g) cobbles, (t) truncations, (b) burins, (e) endscrapers, (a) Aurignacian pieces, (r) miscellaneous retouched pieces, (u) utilized pieces, (n) notched pieces. Circumference of circle represents class maximum value of 100.

maximum within-class values for several classes. Debitage, miscellaneous retouched pieces, and notched pieces are well represented in this zone. Zones 1, 5, and 7 are the only medium scale accumulations showing high within-class percentage values for all ten artifact classes. Other zones are composed of more limited artifact sets. Zone 6, to the north of the hearth, has middle range standardization values for truncations, utilized pieces, and notched pieces. Zone 2 has a middle range value for cobbles. Other zones have limited quantities of all artifact classes.

Figure 87 illustrates medium scale class maximum standardizations. Three dense cluster zones are indicated by large polygons, and there are some differences among these. Zones 1, 5, and 7 all have high values for utilized and miscellaneous retouched pieces. Zone 1 differs from the other two in having peaks for cores and burins. Zone 5, the hearth area, has high values for most classes, but cores, endscrapers, and burins are not well represented. Zone 7 is similar to zones 1 and 5 in most respects. However, high values for endscrapers and low values for cobbles and Aurignacian pieces differentiate this zone from others on the surface. Zone 6, although sparse, has a class maximum profile similar to that for zone 7. There are, however, relatively few truncations and less debitage. Zones 6 and 7 correspond to deposits associated but not centered on the hearth in grid unit A1. That they have similar profiles here is consistent with an interpretation of them as accumulations of material generated at the feature and dumped a short distance away. We will return to these similarities shortly.

The sparse zones defined at the medium scale are quite variable in their class maximum profiles. For zones 2, 3, and 8, utilized pieces are well represented in the local assemblage. Debitage is present in both 3 and 8, while notched pieces link 2 and 3. Zone 2 is the only one of these sparse zones containing cobbles. Still, for all of these zones, artifacts from all classes are present in small quantities. In terms of relative abundance, these sparse zones seem to be similar in structure (if not content) to the peripheral zones at Pincevent.

Table 33 gives the zone maximum standardization matrix for the VIII-1 medium scale. Again, debitage is eliminated from consideration because of its overwhelming numerical superiority in all zones. Figure 88 displays these values graphically for each zone.

The three richest zones in the northern sector are similar in content. Zones 6 and 7, located to the north and west of the hearth, have nearly identical profiles. Zone 6 contains cobbles that 7 does not. Zone 7 contains truncations. Zone 5, the hearth, contains few cores, many truncated pieces, and no endscrapers in contrast to zones 6 and 7. A high value for Aurignacian pieces and a low

TABLE 32.

Matrix of class maximum standardization values for Le Flageolet I level VIII-1 artifact classes when eight cluster zones are considered. All values are rounded to the nearest integer.

MATERIAL CLASS	CLUSTER ZONE 1	2	3	4	5	6	7	8
Cores	100	0	0	0	30	60	50	0
Cobbles	0	60	0	0	100	50	0	0
Truncations	45	0	0	9	100	0	55	0
Burins	100	0	0	18	45	9	18	0
Endscrapers	55	0	0	45	0	45	100	0
Aurignacian	55	0	0	36	100	9	27	9
Retouched	100	0	0	13	88	38	88	13
Utilized	100	25	17	0	100	50	67	33
Notched	43	14	43	0	100	57	71	0
Debitage	55	0	9	0	72	26	100	12

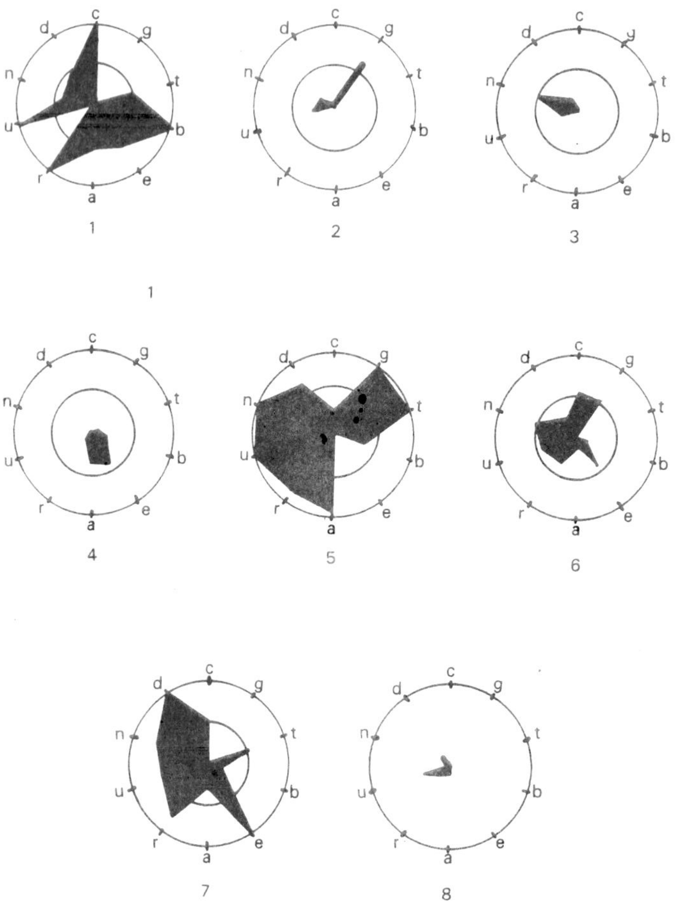

Figure 87. Circular Graphs for Class Maximum Standardization for Zone Contents when Eight Zones Are Considered for Level VIII-1. Radii correspond to artifact classes as follows: (c) cores, (g) cobbles, (t) truncations, (b) burins, (e) endscrapers, (a) Aurignacian pieces, (r) miscellaneous retouched pieces, (u) utilized pieces, (n) notched pieces, (d) debitage. Circumference of circle represents class maximum value of 100.

TABLE 33.

Matrix of zone maximum standardization values for Le Flageolet I level VIII-1 artifact classes when eight cluster zones are considered. All values are rounded to the nearest integer. Note that debitage has been excluded from the calculations.

MATERIAL CLASS	CLUSTER ZONE 1	2	3	4	5	6	7	8
Cores	83	0	0	0	25	100	45	0
Cobbles	0	100	0	0	83	83	0	0
Truncations	42	0	0	20	92	0	55	0
Burins	92	0	0	40	42	17	18	0
Endscrapers	50	0	0	100	0	83	100	0
Aurignacian	50	0	0	80	92	17	27	25
Retouched	67	0	0	20	58	50	64	25
Utilized	100	50	67	0	100	100	73	100
Notched	25	17	100	0	58	67	45	0

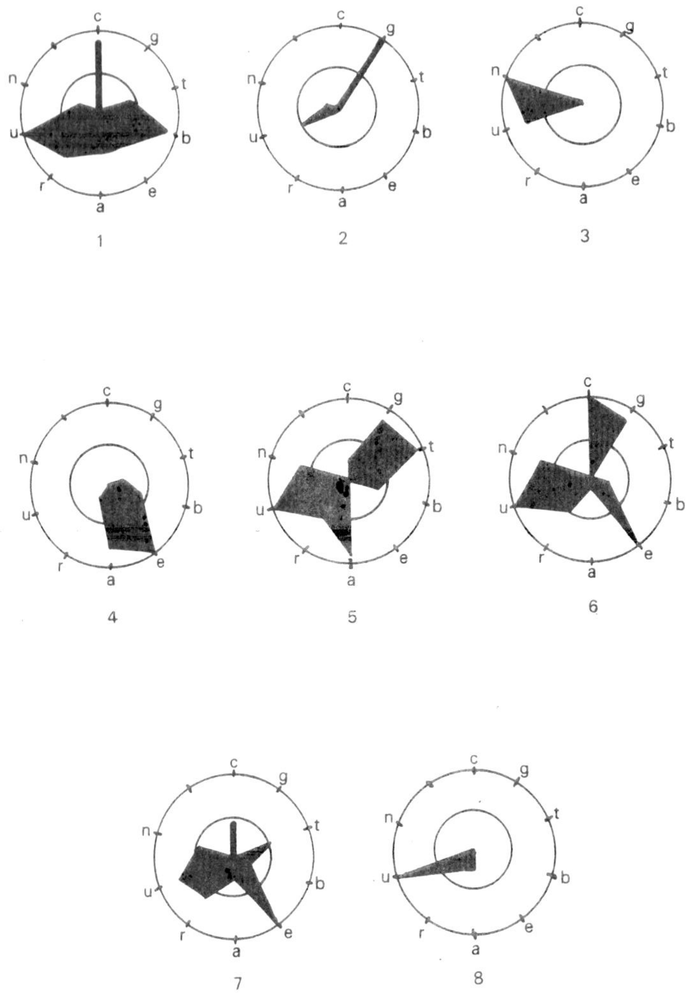

Figure 88. Circular Graphs for Zone Maximum Standardization of Zone Contents when Eight Zones Are Considered for Level VIII-1. Radii correspond to artifact classes as follows: (c) cores, (g) cobbles, (t) truncations, (b) burins, (e) endscrapers, (a) Aurignacian pieces, (r) miscellaneous retouched pieces, (u) utilized pieces, (n) notched pieces. Circumference of circle represents class maximum value of 100.

value for cores differentiate the hearth area from other rich northern zones. The fourth north sector zone, 8, is a sparse accumulation of utilized pieces along with a few Aurignacian and miscellaneous retouched pieces.

In the southern zone, half of the zone maximum profile resembles that for rich northern zones; utilized pieces, miscellaneous retouched pieces, and notched pieces are present in similar proportions. However, other aspects of zone 1 are different. Burins are well represented and make up a large percentage of the local assemblage. Both truncations and endscrapers are present in zone 1 while these are separated in the north. Cores are abundant in zone 1.

The three sparse middle sector zones are diverse. Zones 2 and 3 have notches and utilized pieces in common, although the relative contribution of each to zone assemblages is reversed. Zone 2 is dominated by cobbles which are absent from zone 3. Zone 4, closest to the hearth of all middle sector zones, is composed primarily of tools with endscrapers the dominant artifact form. Truncations, burins, and Aurignacian pieces are also present in zone 4 in reduced proportions. Retouched pieces complete the zone 4 assemblage profile.

Five varieties of zone can be distinguished at the medium scale based on standarized frequency counts. One type centers on the hearth, although this zone is similar to nearby accumulations in many ways. Two dense concentrations to the north and west of the hearth together compose a second variety. Both of these zone types share similar profiles for notched, utilized, and miscellaneous retouched pieces. The hearth zone differs from the zones to its north in other ways. Because of its similarity to hearth area deposits, the second zone variety may represent hearth-related dumps. The southern sector zone (zone 1) represents a third content variety because of high core and burin profiles. Zones 2, 3, and 8 are collected as a fourth variety, characterized by utilized pieces as the dominant assemblage component. Finally, zone 4 (with a high endscraper profile and predominantly tool assemblage) is distinguished as the fifth zone variety.

Although varied deposit characteristics obtain in the northern and southern sectors of level VIII-1, interpretation of the southern sector must await analysis of high scale solutions. It is tempting to place zone 1 with other zone varieties under the rubric "heterogeneous" because of its divergent content. However, remembering that further spatial refinement occurs between scales primarily in the southern sector, such an assignment would be premature. Still, the difference between northern and southern sector zones noted at the low scale is amplified at this medium solution level.

Table 34 lists the class maximum standardized values for twelve zones (the level VIII-1 high scale solution). A clear discrepancy between the northern and southern zones is now obvious, since all of the lower numbers appear in the south. Only one artifact class, burins, has a high value in a southern sector zone at this scale. Burins are also well represented in zone 3. Zones 3 and 4 are the richest in the southern sector and are located at the northern end of the large block that seems to be the focus of deposition in this sector. Utilized pieces have middle range percentages in these two zones. A complicated set of five zones is indicated for the northern sector of the surface.

Figure 89 shows the circular graphs for Table 34. The similarity of southern zones 3 and 4 is clear. Only a lack of notched pieces and the presence of cobbles in zone 4 differentiate the two profiles. Zone 2 can now be related to 3 and 4 based on a relatively high percentage of burins in the local assemblage. Only zone 1 of all zones around the large block shows a distinctive profile. However, the dominant elements in all other southern sector zones (utilized pieces and burins) are well represented in zone 1. Also part of the southern complex, zone 5 lacks cobbles but is similar to the others in most respects.

The northern zone complex displays some internal variety at this highest solution scale. The formation of a fifth zone removes some of the consistency in class maximum profiles noted at the preceeding scale. Zones 8 and 11 are most similar here even though there are important differences between the two. Zones 9 and 11 are also similar in some respects, but the dominant classes in each are different. In fact, the four rich northern zones (8, 9, 10 and 11) seem to represent a continuum in content variety. This continuum may be broken when zone maximum standardization is considered. Still, zone 8, corresponding to accumulations centered on the hearth and a linear scatter of materials away from the hearth, contains the maximum values for four artifact classes and high values for two others. Interestingly, burins are present in zone 8.

Sparse zones include 6, 7, and 12. These are located in the middle and northern sector of level VIII-1. At this clustering scale, no similarities are apparant among these zones; each contains a limited local assemblage composed of few artifact classes. Such variation suggests that sparse accumulations resulted from specific activities. Two of the sparse zones occur in the middle sector where they may reflect stochastic deposition, e.g., loss by dropping. These may also have resulted from limited and infrequent activities carried out on the surfaces of rocks. However, given the topography, it does not seem likely that sustained or repeated activities would have been carried out in the middle sector.

TABLE 34.

Matrix of class maximum standardization values for Le Flageolet I level VIII-1 artifact classes when twelve cluster zones are considered. All values are rounded to the nearest integer.

ARTIFACT CLASS	CLUSTER ZONE 1	2	3	4	5	6	7	8	9	10	11	12
Cores	0	33	50	67	0	0	0	0	33	83	100	0
Cobbles	0	0	0	50	0	50	0	100	67	0	83	0
Truncations	33	0	17	17	0	0	17	100	33	100	50	0
Burins	25	50	75	100	25	0	50	75	50	50	25	0
Endscrapers	0	10	10	40	0	20	0	30	100	0	50	10
Aurignacian	0	13	25	25	13	0	38	0	100	38	50	13
Retouched	0	0	29	43	29	0	0	71	43	100	28	14
Utilized	38	13	50	63	25	13	0	100	75	100	50	50
Notched	0	29	14	0	0	43	0	100	0	71	57	0
Debitage	0	7	13	21	15	8	0	29	46	100	23	12

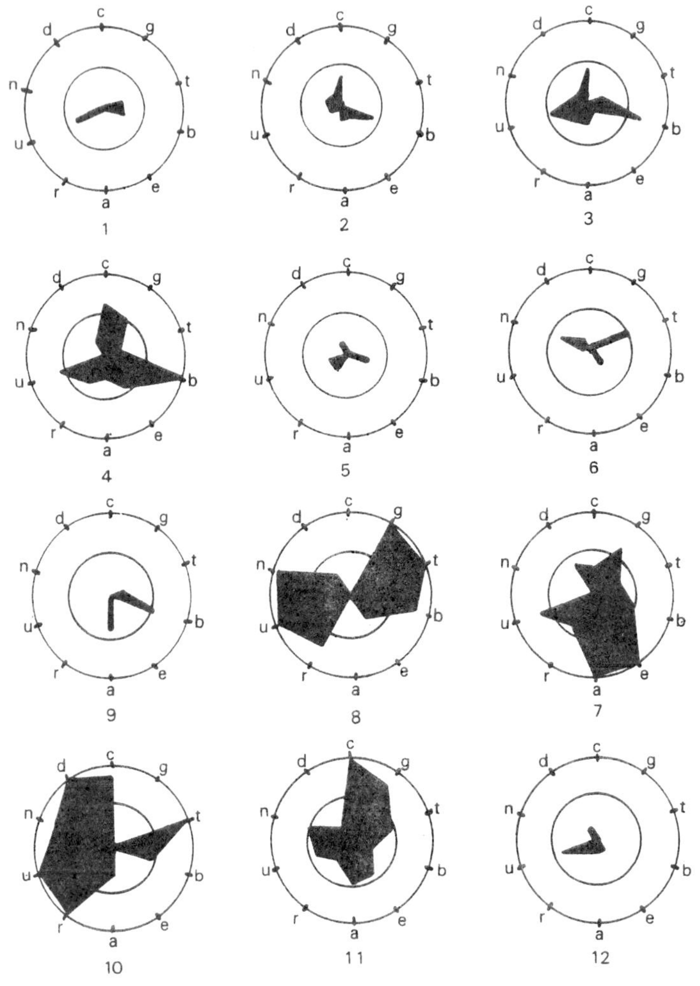

Figure 89. Circular Graphs for Class Maximum Standardization of Zone Contents when Twelve Zones Are Considered for Level VIII-1. Radii correspond to artifact classes as follows: (c) cores, (g) cobbles, (t) truncations, (b) burins, (e) endscrapers, (a) Aurignacian pieces, (r) miscellaneous retouched pieces, (u) utilized pieces, (n) notched pieces, (d) debitage. Circumference of circle represents class maximum value of 100.

Table 35 is the matrix of zone maximum standardization values for the level VIII-1 high scale solution. Again, three southern sector zones show similar profiles (2, 3, and 4). These zones are characterized by high values for cores, burins, and utilized pieces. Some variation exists for other artifact classes, i.e., notches, miscellaneous retouched pieces and endscrapers. Figure 90 shows the similarity in southern sector zones. Cluster zone 5 is similar to this group in its essential qualities, lacking only cores. Zone 1, a sparse concentration at the south edge of the excavated area, is distinct from other southern sector zones.

A more obvious pattern of zone content emerges in the northern sector when zone maximum values are considered. Two content varieties can be distinguished among the four richest accumulations. As was the case at the preceeding scale, the hearth area (zone 8) is similar in content to the accumulation centered in C0/D0 (zone 10). Both zones have many notched, retouched, and utilized pieces. Similar profiles occur for burins and truncations. Cobbles dominate the hearth area and are absent in zone 10. A lack of endscrapers in zone 10 contrasts with results at the previous clustering scale. Zone 10 contains cores which do not occur in substantial numbers in the hearth area. Variation in zone 10 content between clustering scales is due to the definition of zone 9, which is formed at the highest level by subdividing zone 10. Zone 9, located closer to the hearth than either 10 or 11, contains most of the endscrapers that had been assigned to zone 10 at the preceeding scale. This characteristic, coupled with the high percentage for Aurignacian pieces in the zone 9 assemblage, links it to zone 11. Zone 11 is a linear zone to the north of the hearth along the shelter wall. It was previously interpreted as a hearth-related dump or as an accumulation of materials resulting from hearth maintenance. The presence of cobble fragments, cores, and notched pieces in relatively high proportions differentiates zone 11 from zone 9. If the former resulted from sweeping debris out of the hearth, then the presence of cobbles is reasonable. Other differences must be considered as variation between two similar accumulations. The last northern zone (zone 12) is characterized by the dominance of utilized pieces in an otherwise sparse concentration. Zone 12 shares few characteristics with the other, richer north sector zones.

To summarize the results of standardization procedures at the highest clustering scale, the following characteristics are noted. Two kinds of content patterning exist, and these are separate in the northern and southern sectors. In the north, three local assemblage varieties are defined. One content variety comprises a zone centered on the hearth and a second zone centered in units C0/D0. These zones have assemblages dominated by miscellaneous retouched pieces, pieces showing evident use damage, and notched pieces. This

TABLE 35.

Matrix of zone maximum standardization values for Le Flageolet I level VIII-1 artifact classes when twelve cluster zones are considered. All values are rounded to the nearest integer. Note that debitage has been excluded from calculations.

ARTIFACT CLASS	CLUSTER ZONE 1	2	3	4	5	6	7	8	9	10	11	12
Cores	0	100	75	80	0	0	0	0	20	63	100	0
Cobbles	0	0	0	60	0	100	0	75	40	0	83	0
Truncations	67	0	25	20	0	0	33	75	20	75	50	0
Burins	33	100	75	80	50	0	67	38	20	25	17	0
Endscrapers	0	50	25	80	0	67	0	38	100	0	83	25
Aurignacian	0	50	50	40	50	0	100	0	80	38	67	25
Retouched	0	0	50	60	100	0	0	63	30	88	33	25
Utilized	100	50	100	100	100	67	0	100	60	100	67	100
Notched	0	100	25	0	0	100	0	88	0	63	67	0

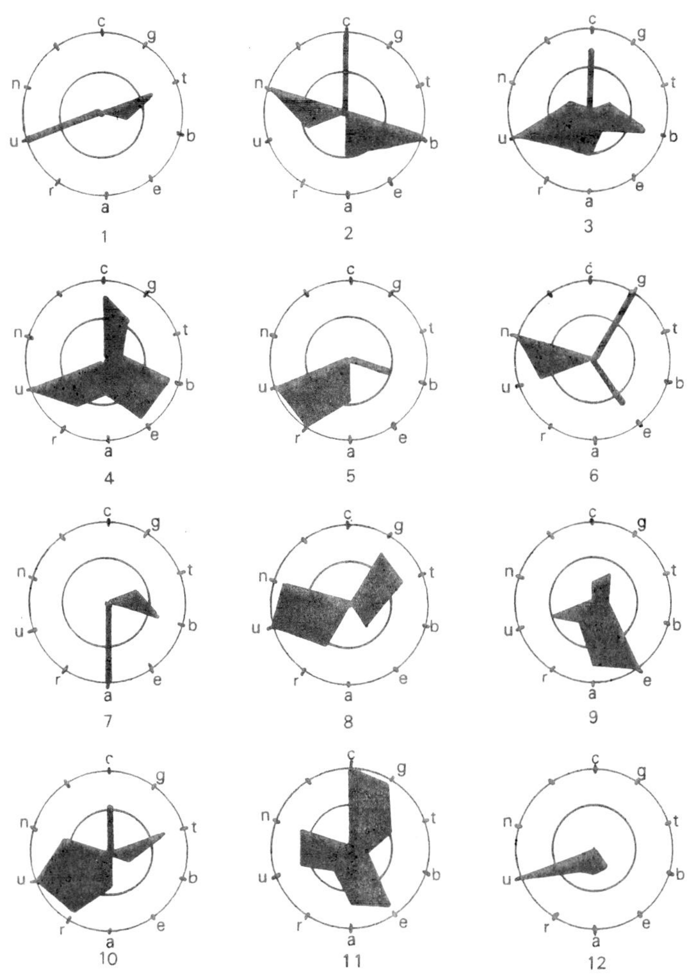

Figure 90. Circular Graphs for Zone Maximum Standardization of Zone Contents when Twelve Zones Are Considered for Level VIII-1. Radii correspond to artifact classes as follows: (c) cores, (g) cobbles, (t) truncations, (b) burins, (e) endscrapers, (a) Aurignacian pieces, (r) miscellaneous retouched pieces, (u) utilized pieces, (n) notched pieces. Circumference of circle represents class maximum value of 100.

similarity suggests that the C0/D0 concentration represents a dump for materials generated in hearth-area activities. The absence of cobbles in C0/D0 supports the notion that these materials were removed from the hearth area as they were generated rather than swept away after initial depostion at the feature. The presence of cores, debitage, and dechet (see Chapter 6) in C0/D0 indicates that the waste from lithic reduction (which might have occurred near the hearth) were disposed of as a unit into this deep crack.

Activities restricted to the hearth area but not removed to a dump include those performed using endscrapers. These artifacts are localized in an accumulation centered in grid unit B0 just to the west of the feature. This zone represents a second variety in the northern sector and is similar to the fourth rich zone, positioned to the north of the hearth along the wall. These contain cobbles and seem to reflect the removal of materials directly from the hearth by sweeping. The third northern sector zone variety is the sparse zone 12; this accumulation is different than all others in the sector.

In the southern sector of level VIII-1, a rather generalized set of deposits follow the edge contours of the large block. Zonation becomes apparant only at the finest scale of resolution, but zone content is similar at the fine level. The southern zones compose another content variety. The structure of zone distributions in the south and their overall similarity in content suggests a rather specialized use of this space during the VIII-1 occupation. Perhaps the large block surface dominating the southern sector was the locus of a specific activity, and tasks were performed repeatedly with the refuse generated from each task deposited into the cracks around the block. Such use of the space would account for the replicated small scale assemblages (comprising burins, cores, and utilized pieces) that are scattered around the block.

Between the two large sectors of the site, small and sparse accumulations, quite variable in content, are defined at the highest scale. In each one, a single artifact class is usually dominant. However, sample sizes are so small that caution is required in interpretation. Stochastic processes (e.g., loss by dropping) may have led to the generation of small scale accumulations, and their significance is uncertain. That they have spatial integrity is clear; to what causes content variation can be attributed is not. Nevertheless, they do represent a distinct and variable kind of material accumulation in level VIII-1.

DISCUSSION AND CONCLUSIONS

A basic pattern in zone content crosses over all three scales of spatial resolution in level VIII-1. Both standardization axes have similar profiles for zones in the southern sector of the occupation over all scales. This indicates an essential unity in content for this sector regardless of spatial resolution. While this might be taken as reflecting homogeneity in the southern sector, it is important to note that the overall profile of the sector is quite distinct from that in northern zones--indeed, from that for the site as a whole. A restricted set of materials is accumulated in the south of level VIII-1. These artifacts were deposited as small concentrations, similar in content, around a large and centrally located block. When all of the southern sector accumulations are combined into a single unit (i.e., at the lowest clustering scale) they have a similar content profile to each individual accumulation defined at the highest resolution scale.

In the northern sector of level VIII-1, no such pattern is evident over solution scales. Around the Al hearth, many varieties of zone are indicated by content analyses. With the exception of burins, all artifact classes have their highest frequencies somewhere in the northern sector. Yet, the zones are different. The accumulation associated directly with the hearth feature is the most general in character with a wide range of classes mixed together in relatively large proportions. Other zones around the hearth area seem to represent related deposits, although each is dominated by (and also lacks) one or a few different artifact categories. A fan-shaped refuse disposal pattern, organizationally similar to those at Pincevent, is present in level VIII-1 but constrained by the distribution of large rocks in the northern sector. It might be that the method of disposal varied around the hearth as much as the content of refuse. One dump is interpreted as reflecting hearth sweepings (the accumulation parallel to the shelter wall north of the hearth). Other deposits seem to reflect removal of materials from the hearth area for disposal into natural "refuse bins". It is clear that knapping debris constitutes one class of removed refuse since all the usual waste products of lithic reduction (exhausted cores, debitage, and dechet) have their highest frequencies in a single zone centered on unit C0.

The most relevant solutions for each sector in level VIII-1 are reached at different points during spatial analysis. The northern sector is characterized first (at the medium scale) while the southern sector is essentially ignored in zone formation until the highest scale. This difference in structure emphasizes the differences between the two sectors already stressed at length.

That the group maintenance focus of level VIII-1 occupation occurred in the northern sector is borne out by spatial analysis. In that area, the only hearth for the level is surrounded by deposits similar to those from the Pincevent "habitation areas". A second, different focus is apparent around (perhaps on) the large block in the south. Low scale solutions in both areas show a similar pattern; cluster centroids for all artifact classes are positioned to the north of each focus. As more clusters are defined by k-means, the hearth continues to be a center for material accumulations. In the south, deposits take on a linear pattern around the edges of the block. Clearly, the two areas do not represent similar structural entities.

A hypothetical interpretive map for the level VIII-1 surface can be generated from the results of spatial analyses. Figure 91 illustrates the overall organization of Le Flageolet during the VIII-1 occupation in terms of general variety in formation processes. A series of dumps and sweeping zones are inferred around the northern sector hearth. The southern sector is interpreted as functionally specific space under the assumption that group maintenance tasks performed around the hearth are "functionally general," i.e., performed at most sites regardless of why people occupied the place. Table 36 describes the varied content of each kind of inferred deposit. The percentage of each artifact class occurring in dumps, hearth sweeping deposits, the hearth itself, functionally specific spaces and sparse accumulations are listed.

These assemblage variants cannot be viewed as "tool kits", but they do provide information on spatial variability in level VIII-1. Burins are an important component of functionally specific areas (in the southern sector). Over 50% of all burins occur in these contexts. Burins also occur in the hearth context, but only 14% of these tools are located there. Lithic reduction waste is concentrated in dumps, with the C0 dump the primary recepticle for this material. Cores are also well represented in dumps as are truncated, utilized, and miscellaneous retouched pieces. The last two artifact classes (along with notched pieces) required little modification for production. They might be considered as "expedient" tools manufactured quickly for immediate use. These expedient tools dominate the hearth area assemblage. However, one tool class requiring more intensive modification for production, endscrapers, is also present in hearth area dumps.

The overall picture of level VIII-1 is of a small-scale, temporary occupation by a single group. The presence of a single hearth, diffuse and only lightly maintained by cleaning, and a series of small, discrete refuse accumulations around the feature do not warrant an inference of intensive occupation. The reservation of

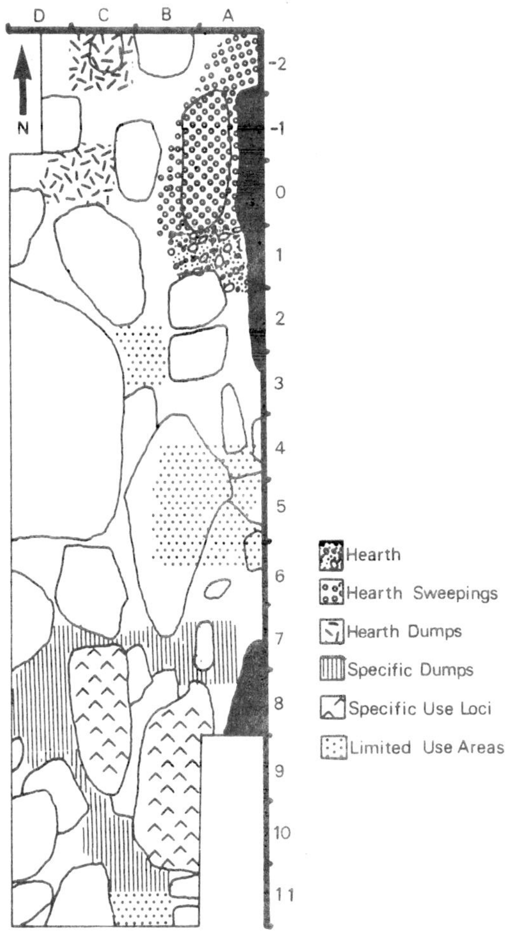

Figure 91. Hypothetical Reconstruction of Level VIII-1 Organization Based on Complete Spatial Analysis.

TABLE 36.

Percentages of artifacts by class contained in each inferred deposit type for Le Flageolet I level VIII-1.

ARTIFACT CLASS	DEPOSIT TYPE HEARTH	SWEEPINGS	DUMPS	SPECIFIC	SPARSE
Cores	0.0	36.4	22.7	40.9	0.0
Cobbles	28.6	42.9	0.0	14.3	14.3
Truncations	27.3	22.3	27.3	18.2	4.5
Burins	15.0	15.0	10.0	50.0	10.0
Endscrapers	11.1	55.6	3.7	22.2	7.4
Aurignacian	12.0	36.0	16.0	24.0	12.0
Retouched	20.0	20.0	32.0	28.0	0.0
Utilized	17.0	21.3	25.5	31.9	4.3
Notched	31.8	18.2	22.7	13.6	13.6
Debitage	1.1	27.8	45.6	22.2	3.3

nearly half the available space for special activities further suggests a functionally specific, short term use of the site. The environmental evidence for local resource exploitation and the raw material evidence for mobility also support this inferred site utilization.

In comparing the results of level VIII-1 analysis to those obtained for Pincevent, similarities and differences can be noted. In one important respect the two sites are similar. Occupations at both sites center material depositon on focal hearth features. Such features have similar spatial organization with refuse accumulations forming an arc to one side of the hearth. At both sites, hearth area accumulations are the richest, and deposits become less rich as distance from the feature increases. At Pincevent, a concentric fan model characterizes refuse deposits, and functionally specific accumulations are rare. In Le Flageolet I level VIII-1, refuse deposits occur in cracks between rocks, and specific accumulations are common at the highest clustering scale, dominating the entire southern sector of the surface. As spatial refinement increases, homogeneity in zone content at Pincevent does not appreciably change. Homogeneity decreases dramatically among level VIII-1 accumulations as finer-scale patterns are discerned.

Pincevent and Le Flageolet I level VIII-1 have different spatial structures. These differences may relate to distinct behaviour patterns among Magdalenian and Aurignacian peoples (a "cultural" explanation) or they may relate to variation in site location, form, and/or topography (a "functional" explanation). If the latter explanation is warranted, then there should be little change in site use at Le Flageolet during the Aurignacian sequence since, as we have seen, there is little change in ecological conditions or topography. It will be more difficult to address the "cultural" explanation because the later levels at Le Flageolet (Upper Perigordian in Le Flageolet I and Magdalenian in Le Flageolet II) have not yet been studied in terms of spatial organization.

Comparisons between Aurignacian occupations from Le Flageolet will still provide information concerning the use of the site over time. That level VIII-2 and level IX are of the same Aurignacian technological tradition controls for variation in that dimension. It remains to be seen if site use at Le Flageolet was similar during other Aurignacian occupations of the shelter and if topographic constraints on distributions operated in the same ways. Was site use redundant? We turn now to the analysis of level VIII-2 to answer this question.

CHAPTER 8

LE FLAGEOLET I: LEVEL VIII-2

INTRODUCTION

The same ten artifact classes used in the analysis of level VIII-1 are employed here (Table 37) All results are thus comparable between the two Le Flageolet Couche VIII levels. An identical sequence of analytic stages is carried out as for the two previous studies, and the same statistics and graphic presentations are used to assess variation in content and artifact association.

Some General Expectations

At Pincevent and in level VIII-1 from Le Flageolet I, certain hearths were the foci of habitation. In both cases, it is the structure of deposits around the features that is of primary utility in identifying a particular hearth as focal. Focal hearths are associated with dense dump zones. The content of these zones reflects the presence of a wide range of artifact classes, both tools and other materials. Focal hearth dumps are dense (often the richest accumulations on a surface) because they reflect the repeated deposition of materials generated during basic and ubiquitous group maintenance activities. Focal hearth areas and associated dumps should occur in most sites regardless of the site's function within a regional economic system; the food preparation, consumption, and general craft activities reflected in hearth area accumulations must occur nearly everywhere in order to sustain human life.

A basic difference between Pincevent and Le Flageolet has to do with the number of focal hearths present. Three focal hearths were identified at Pincevent, and three social units are thus inferred. At Le Flageolet I during the level VIII-1 occupation, only a single such feature is identified, and that is a diffuse and sparse example. A single social unit is therefore inferred. It should be noted that no equivalence is assumed among these "social units". Their nature and composition, and variability between them, remains unknown and will not be inferred.

These considerations are important when the contextual elements uncovered in level VIII-2 are recalled. These include a hearth area located in the northern sector of the surface and possibly a second, smaller feature in the south on the surface of a large

TABLE 37.

Frequencies of ten artifact categories analyzed for Le Flageolet I level VIII-2 assemblage.

ARTIFACT CLASS	N
Endscrapers	34
Burins	40
Aurignacian Pieces	19
Truncated Pieces	20
Notched and Denticulated Pieces	41
Miscellaneous Retouched Pieces	46
Utilized Pieces	71
Cores	25
Debitage	750
Cobbles and Fragments	213

block. Materials recorded by meter square provenience suggest a "focal" pattern for the northern hearth. All categories of small artifacts have overlapping distributions in a fan-shaped pattern to the north and west of the feature. Similar overlapping distributions occur in the southern sector below the burnt area of the B7-B10 block; these are less rich than northern sector accumulations.

Two scenarios can be imagined that conform to the level VIII-2 pattern as indicated in Chapter 6. The first involves the presence of two focal hearths, and thus two social units. Under this hypothesis, homogeneity of accumulations in the north and south sectors of the surface should be the same throughout the refinement of spatial patterning. Variation in zonation must be accounted for by variability in available space (topographic constraints).

If it cannot be demonstrated that the two sectors are structurally the same, then a second hypothesis must be invoked. If the two sectors of level VIII-2 have distinct organizations (as was the case for level VIII-1), then even the presence of two hearth features cannot be used to infer the presence of two social units during the occupation. A similar interpretation of spatial organization must be offered for this situation as for level VIII-1, although special functions in the southern sector might have involved the use of fire in level VIII-2. Under this hypothesis, variability between the northern and southern sectors is seen as related to the organization of activities over space. Around the focal hearth, a series of dump accumulations is expected. More restricted sets of artifacts are expected in more functionally-specific areas. Patterns such as these would indicate a similar kind of occupation to that inferred for level VIII-1, but the activities performed in specialized areas need not have been the same. Even if they were not, organizational redundancy in site use would be indicated.

STAGE I ANALYSIS: PATTERN RECOGNITION

Each of the ten level VIII-2 artifact class distributions was entered into k-means clustering runs. As was the case for all previous analyses, clustering was carried out until fifteen clusters were formed by the procedure. Figure 92 shows the log(%SSE) plots produced by these runs for all ten artifact classes. As was the case for level VIII-1, all classes have optimal solutions at the two cluster level. Three classes (cobbles, miscellaneous retouched pieces, and debitage) have only two optimal solutions. Five categories have three optimal solutions indicated by the log(%SSE) plot: truncations, endscrapers, Aurignacian pieces, utilized pieces, and notched and denticulated pieces. Four of these terminate

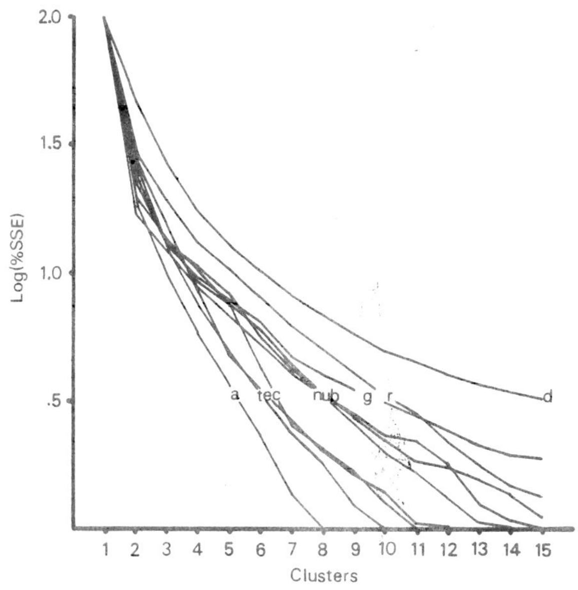

Figure 92. Plot of log(%SSE) Statistics for All Ten Artifact Classes in Le Flageolet Level VIII-2 as Defined by K-means Cluster Analysis to Fifteen Clusters.

TABLE 38.

Optimal clustering solution levels indicated by log(%SSE) plots produced by k-means analyses of Le Flageolet I level VIII-2 artifact distributions. A "t" indicates a terminal solution at zero before fifteen clusters are formed.

ARTIFACT CLASS	SOLUTION NUMBER 1	2	3	4
Cores	2	5	10	12t
Cobbles	2	7		
Truncations	2	9	10t	
Burins	2	7	13	15t
Endscrapers	2	5	11t	
Aurignacian Pieces	2	7	8t	
Retouched Pieces	2	4		
Utilized Pieces	2	6	10	
Notches and Denticulates	2	7	14t	
Debitage	2	9		

at zero before fifteen clusters can be formed. Two classes have fourth solutions indicated, both terminal: cores and burins. Table 38 lists the optimal solution levels for all artifact classes as determined by log(%SSE) plots.

Two general patterns of cluster formation in the level VIII-2 class distributions are apparant. One pattern begins with two clusters, passes through a medium solution where the two inital clusters are divided into equal numbers of clusters, and ends with a third level where the two lowest scale clusters are still equally divided. Classes having this clustering pattern include debitage, cores, retouched pieces, notched pieces, cobbles, and utilized pieces. Notched pieces provide a good example of this sequence. Figure 93 shows the map for notched pieces produced from k-means statistics at the lowest clustering scale (two clusters). The two clusters are located in the northern and southern sectors with one centroid in grid unit C7 and another in B0. The southern cluster centers on a grid unit with dense, overlapping concentrations of small artifactual material. Figure 94 shows the same distribution divided into seven clusters. Three clusters have been defined in each sector with a single isolated piece located at the extreme southern end of the excavated area. In the northern sector, clusters are centered north and west of the hearth. In the south, clusters occur on either side of the burnt block just below the area of burning. Two clusters are positioned to the west of the block. Figure 95 shows the (termination) fourteen cluster solution for notched pieces. Here, four clusters and two isolates are located in the southern sector. Eight clusters are positioned in a band running to the north and west of the hearth.

The second kind of clustering sequence noted for level VIII-2 artifact classes begins with a two cluster solution but differs from the first pattern at the medium and high scales. At medium levels, many more clusters are formed in the north sector than in the south. This trend continues at the highest level, where the northern sector is further subdivided. This pattern group contains truncated pieces, burins, endscrapers, and Aurignacian pieces. Figure 96 shows the two cluster solution for endscrapers. This map is similar to that for notched pieces. One cluster centroid is located in each sector of the surface, centered in grid units C7 and B0. A five cluster pattern provides a medium solution for the endscraper distribution, and three clusters are formed in the northern sector while two are located in the south (Figure 97). In the north, one centroid is positioned at the A1 hearth, a second to the west in unit D2, and the third in unit B-1. Southern cluster centroids are located in units C7 and C10. At the highest clustering scale for endscrapers, eleven clusters are formed. Eight of these clusters are located in the northern sector while only three are in the south (Figure 98) In the south, one of

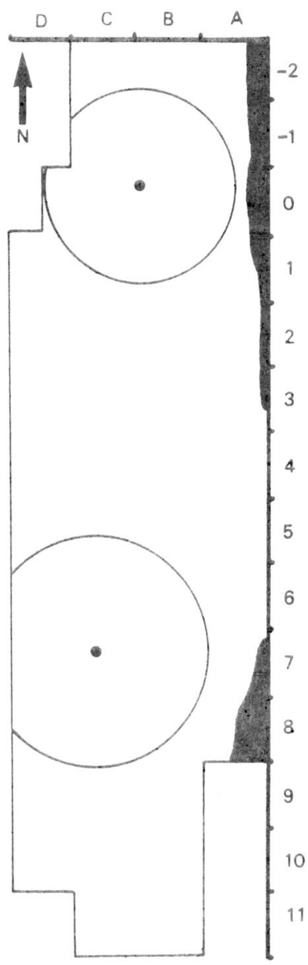

Figure 93. Map of Two Cluster K-means Solution for Le Flageolet I Level VIII-2 Notched Pieces Distribution. Dots mark cluster centroids; circles are RMS radii for clusters.

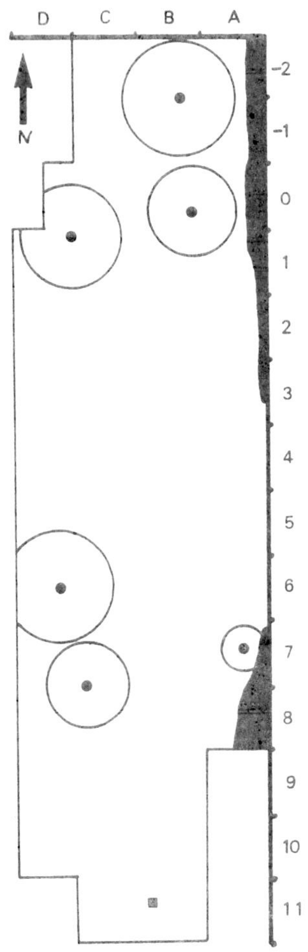

Figure 94. Map of Seven Cluster *K*-means Solution for Le Flageolet I Level VIII-2 Notched Pieces Distribution. Dots mark cluster centroids; circles are RMS radii for clusters.

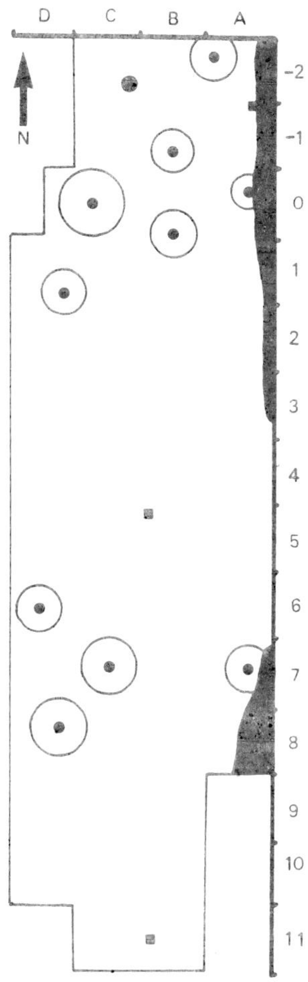

Figure 95. Map of Fourteen Cluster K-means Solution for Le Flageolet I Level VIII-2 Notched Pieces Distribution. Dots mark cluster centroids; circles are RMS radii for clusters.

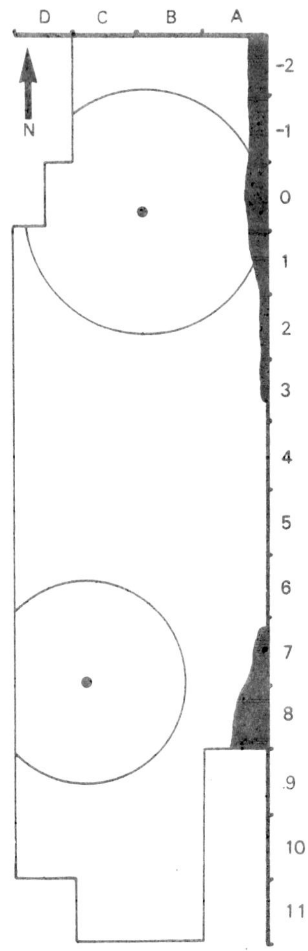

Figure 96. Map of Two Cluster K-means Solution for Le Flageolet I Level VIII-2 Endscraper Distribution. Dots mark cluster centroids; circles are RMS radii for clusters.

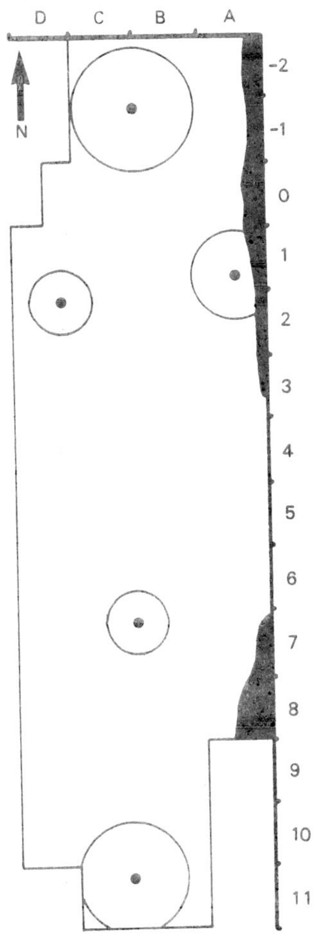

Figure 97. Map of Five Cluster K-means Solution for Le Flageolet I Level VIII-2 Endscraper Distribution. Dots mark cluster centroids; circles are RMS radii for clusters.

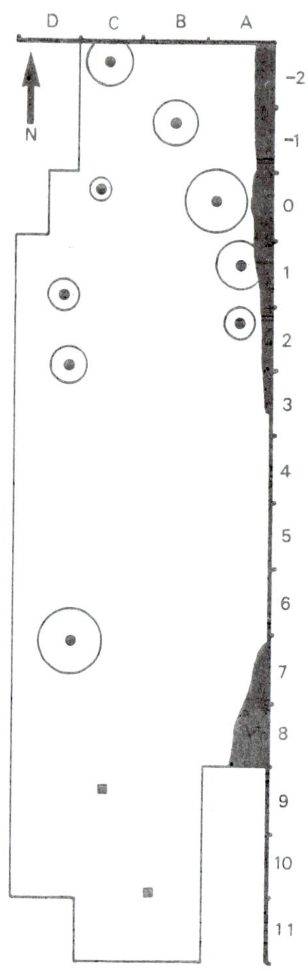

Figure 98. Map of Eleven Cluster K-means Solution for Le Flageolet I Level VIII-2 Endscraper Distribution. Dots mark cluster centroids; circles are RMS radii for clusters.

the medium scale clusters is divided into two isolated pieces; the other cluster remains unchanged. In the north, all clusters have short RMS radii and form two linear patterns from grid unit A2 to C-2 and from D2 to D1.

It should be noted that the artifact classes composing these two pattern groups conform to the distinction made earlier between "expedient" and "labor-intensive" tools. The first pattern (equal division between sectors) obtains for expedient tools. The second pattern (northern sector dominance) characterizes those artifact forms requiring more effort in production. We will return to these patterns when content analyses are discussed.

The spatial refinement exhibited in the level VIII-2 northern sector is similar to that observed in focal hearth areas at Pincevent. In all cases, individual artifact distributions are divided into more clusters from level to level. This tendency has been related to patterns in overall material abundance in previous analyses, and this may also be the case here in level VIII-2. In fact, most artifact classes are twice as frequent in the northern than in the southern sector. The relation between clustering and abundance noted at Pincevent would seem to hold for level VIII-2; the areas of greatest material accumulation are divided first and most often throughout the clustering run.

General characterisitcs of all individual runs include a concentration of clusters around the northern end of the burnt block surface (in unit B7). Specifically, grid units C7 and D7 are the typical loci of centroids. Other cluster centroids are scattered across the southern sector. More detailed discussion of cluster positions must await overlay mapping and zone formation.

The northern sector of level VIII-2 is characterized by a variety of clustering patterns. Endscrapers, for example, have a linear centroid configuration at the high clustering scale. Other classes, e.g., cores, have dispersed patterns as more clusters are defined during the analytic sequence.

Even more than was the case for level VIII-1, there is marked variation here in the number of optimal solutions identified for an artifact class distribution. There is also considerable variety in the number of clusters composing middle and high scale solutions. Most artifact class distributions have three optimal solutions indicated by log(%SSE) plots. Some classes have a fourth, terminal solution defined. Three classes have only two solutions. The latter situation provides the biggest problem in producing overlay maps, since relating solutions to other classes with more than two patterns is difficult. These problems will treated as was the distribution of Aurignacian pieces in level VIII-1. The lowest clustering level for every class will be compared. The second solution for all classes will be overlaid at a medium scale. Third solutions will be used at the highest scale even for classes having a fourth, terminal solution (because preliminary examinations showed no major differences in third and fourth solutions). For classes having only two solutions, the second level will be overlaid at the highest scale with other classes.

With these concessions to consistency in mind, three clustering scales will be observed in the zone definition phase that follows. Again, the three overlay involved maps will be referred to as low, medium, and high solution scales. As was the case for level VIII-1, there will be variation in the number of clusters mapped for different classes at a given scale.

In the following analyses, particular attention will be paid to variability in patterning between expedient and labor-intensive tools. When considered as individual classes, the first group shows scattered cluster configurations equally divided between northern and southern sectors. The second group seems to have more clusters in the northern sector near the hearth, and the number of clusters defined in the north increases throughout the clustering sequence. These general groups also have spatial relevance in level VIII-1; thus, they deserve special attention here.

STAGE II ANALYSIS: PATTERN RECOGNITION, ZONE DEFINITION

To compare the distribution characteristics resulting from k-means analysis of individual artifact classes, three overlay maps are produced. As was the case in level VIII-1, the lowest clustering scale for level VIII-2 is composed of two cluster solutions for all classes (Figure 99). A wider range of appropriate solutions is present at both the medium and highest solution scale. Figure 100 illustrates the medium scale overlay, and Figure 101 shows the highest scale.

Two cluster solutions show a pattern similar to that noted in level VIII-1. Compact masses of centroids are located in the north and south sectors. Cluster centroids for cobbles are located to the east (shelter wall) side of each of these masses. The low scale solution for level VIII-2 differs slightly from the level above in that southern sector centroids are more concentrated here. Other than this minor divergence, the low scale map for level VIII-2 is almost identical to that obtained for the upper Aurignacian surface.

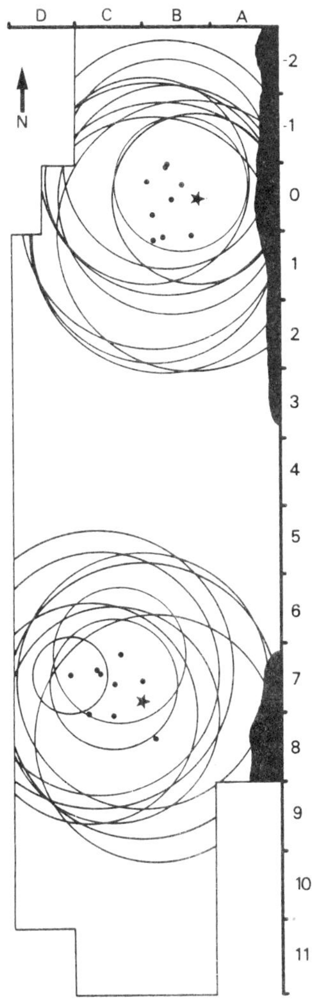

Figure 99. Low Scale Overlay Map Produced by K-means Analyses of Ten Artifact Class Distributions from Le Flageolet I Level VIII-2. Dots mark cluster centroids; boxes mark the position of isolated pieces; circles around centroids represent RMS radii for the clusters; cobble cluster centroids are marked with a star.

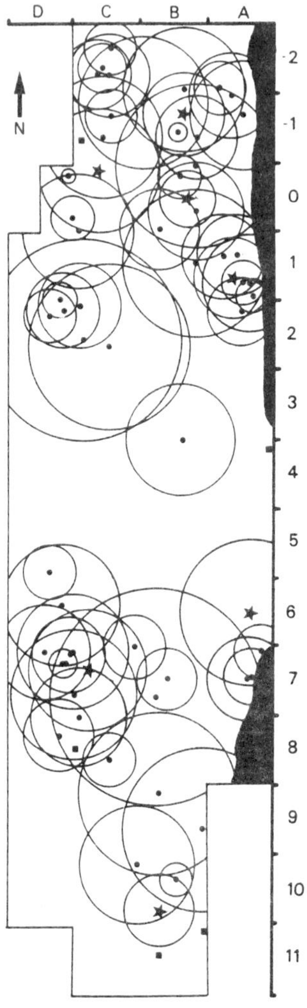

Figure 100. Medium Scale Overlay Map Produced by K-means Analyses of Ten Artifact Class Distributions from Le Flageolet I Level VIII-2. Dots mark cluster centroids; boxes mark the position of isolated pieces; circles around centroids represent RMS radii for the clusters; cobble cluster centroids are marked with a star.

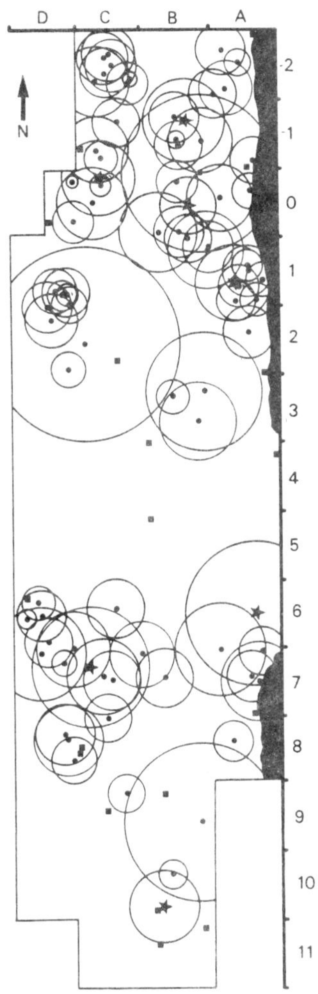

Figure 101. High Scale Overlay Map Produced by K-means Analyses of Ten Artifact Class Distributions from Le Flageolet I Level VIII-2. Dots mark cluster centroids; boxes mark the position of isolated pieces; circles around centroids represent RMS radii for the clusters; cobble cluster centroids are marked with a star.

The northern centroid group is concentrated in grid unit B0 with three centroids located in unit B1 at its boundary with B0. All cluster centroids are contained by all RMS radii. The northern concentration is positioned very near the inferred hearth for level VIII-2 which lies to the southeast in unit A1. The northern cluster group coincides with accumulations of small artifacts interpreted as hearth sweeping deposits in Chapter 6.

The focus of the southern centroid group is in grid unit C7 with single cluster centroids located in D7 and B8. Grid unit C7 is adjacent to the burnt rock surface. The unit also contains quantities of bone, debitage, cobble fragments and dechet.

At the lowest clustering scale, then, similar patterns are observed in both sectors of the surface. Centroid concentrations occur in grid units having many small artifacts of all kinds. Cobble cluster centroids are located to the interior side of all other artifact classes in both sectors. In level VIII-1 a difference between the northern and southern sectors was evident even at the lowest scale, but no such variability can be detected based on centroid patterns here in level VIII-2.

Two zones are defined for level VIII-2 at the low scale of resolution (Figure 102). Southern zone 1 contains clusters for each of the ten artifact classes. Zone 2 in the north, associated with the inferred focal hearth, also has each artifact class represented in its assemblage. Again, no difference between the two sectors can be detected. It should be pointed out, moreover, that topographic conditions vary between the two sectors. In the southern sector, cluster centroids are located in a wide crack between rocks. The locus of northern cluster concentration is a flat rock surface, and the northern block surfaces are only a few centimeters above the surrounding Couche VIII sediment matrix. As has been discussed, it may be that southern materials were also generated on block surfaces and were subsequently deposited beside the rocks. For the moment, zone 2 can be designated as a hearth area zone, but zone 1 cannot firmly be associated with any feature.

The medium scale overlay map for level VIII-2 is shown in Figure 100. As can be seen, the two sectors remain separate at this resolution scale. Within each sector, a complex pattern is now apparant. In the southern sector, a mass of centroids with overlapping RMS radii forms a band from unit D7 into units C8/D8. All artifact classes (including cobbles) are represented in this band. On the other side of the burnt block, four cluster centroids are positioned near the shelter wall. A cobble cluster is present among this group, but it has a very long RMS radius and contains a few dispersed pieces. Finally, a scatter of centroids is located to the south of the large southern rock; their wide RMS circles also

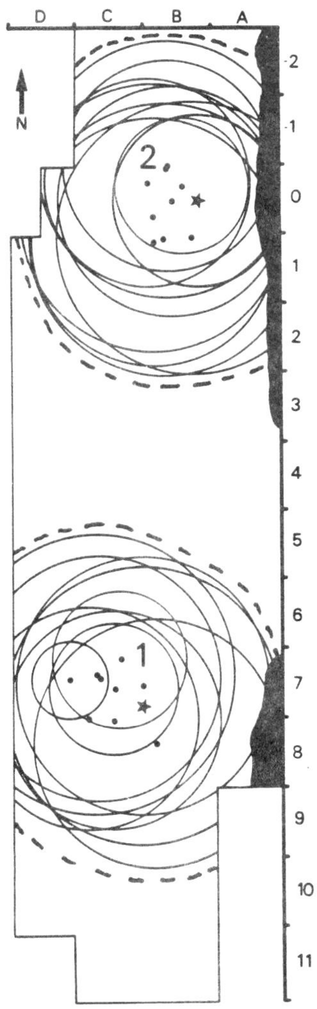

Figure 102. Two Zones Defined at Low Scale Solution for Ten Artifact Class Distributions from Le Flageolet I Level VIII-2. Zones are outlined with a dotted line.

represent dispersed and sparse accumulations.

In the northern sector of level VIII-2, cluster centroids occur over the entire available surface and occupy all cracks between rocks. This is in contrast to the level VIII-1 surface, where northern clusters were concentrated in "pockets" for the most part. Northern clusters here are compact, although a few wide RMS circles are present. These larger clusters are located in the northern and westernmost parts of the excavation.

A single burin cluster is present in the middle sector at this solution scale. That the middle sector is similar between Couche VIII surfaces is unsurprising since topographic conditions changed little between occupations.

Zone definition is relatively easy at the medium scale (Figure 103). Three southern sector zones are defined as discussed above. Zone 1 is the southernmost collection of diffuse clusters. Zone 2 is the group of clusters near the shelter wall. Zone 3 is the band of clusters composing the densest configuration in the southern sector.

Northern zone definition is less evident than in the southern sector. However, there do seem to be six concentrations of cluster centroids and RMS overlaps. Zone 4 is a group of eight clusters, including one of cobbles, centered on the Al hearth. There is little RMS overlap between this zone and neighboring zones 5 and 6. Zone 5 is also a compact group of centroids. The zone contains six clusters, five with short RMS radii, but there are no cobbles concentrated here. A dispersed cluster is assigned to zone 5 at this scale. Even though its RMS radius overlaps with some members of zone 4, the centroid of this cluster is closer to the others in zone 5 than to those in zone 4. Zone 5 centroids are concentrated in a crack between rocks located in grid units D2/D3. Zone 6 has a linear centroid pattern and passes from the Al hearth north into grid units A-2 and B-2. The zone comprises five clusters, one of cobbles. A primary consideration in distinguishing zone 6 as a separate unit is the configuration of surrounding zones. Zone 4 to the south is very compact. Zone 8, to the north near the shelter wall, is also well defined.

Zones 7, 8, and 9 are a variable group of units that complete the northern sector complex at the medium scale. Zone 7 is a linear configuration of six clusters (one for cobbles) confined between blocks in grid units D0 and D-1. Although much of D-1 remains to be excavated, most of the grid unit is occupied by a large block at the level of the VIII-2 deposits. It seems likely that the form of zone 7 results more from internal shelter topography than from excavation extent. Zone 8, as was mentioned above, is a compact set of eight clusters, including one of cobbles, at the extreme northeastern part of the level VIII-2

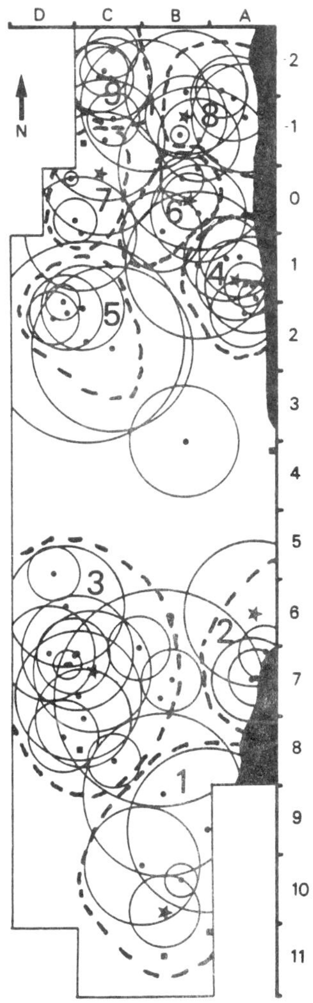

Figure 103. Nine Zones Defined at Medium Scale Solution for Ten Artifact Class Distributions from Le Flageolet I Level VIII-2. Zones are outlined with a dotted line.

surface. The space containing zone 8 is relatively flat, and for this reason centroids have a circular configuration. Finally, zone 9 is a tightly massed group of clusters containing no cobble concentration and located at the extreme northwest corner of the surface.

The results of medium scale zone definition are both similar and different from those attained for level VIII-1. To recall the VIII-1 patterns, northern sector zones were compact and spatially discrete, while those in the south showed no zonation until the highest clustering scale was considered. The reverse is true in level VIII-2. It is the northern sector here that presents difficulties in zone definition at the medium scale, although there is patterning evident. The two levels are similar in that compact and dense zones characterize the hearth areas, and a series of accumulations are arrayed around the hearths, constrained in their shape by the local topography. Higher levels of resolution may clarify zonation in the northern sector, but the basic contrast in configuration between level VIII-1 and level VIII-2 is pointed out now so that any resulting changes can be compared in discussing the highest scale.

Figure 101 shows the level VIII-2 high scale solutions overlaid for all ten artifact classes. Three of these classes, including cobbles, retain their medium scale solution at this highest scale. There is only slight change between medium and high scale patterns. However, as was hoped, northern sector refinement occurs. The southern sector undergoes little modification from the preceeding scale.

Figure 104 shows the eleven zones defined based on Figure 101. Again, three zones are defined in the southern sector, and these are positioned around the large central block. Zone 1 is again the group of clusters at the extreme south of the excavated area. It comprises four clusters (one of cobbles) and three isolated pieces. Three of the clusters are fairly compact. The zone is confined to an open space bounded by a rock at the north and by the edges of excavation at the east and south. Zone 2 corresponds to the medium scale zone of the same number. Located between the block and the shelter wall, six clusters and an isolate compose the zone. One of the clusters is cobbles. Zone 2 has a linear centroid configuration because of its topographic situation. There is very slight overlap of RMS radii between zones 1 and 2. Only a single overlap links zone 2 with zone 3. More importantly, the large block separates the last pair of zones. Zone 3 is the third southern zone. At this scale, zone 3 is a dense mass of clusters on the western side of the central rock and is bounded at its western side by excavation walls. Seventeen clusters and isolates make up the zone. These materials occur over a relatively flat space in grid units D6, C7, D7, C8, and D8. A cobble cluster is present in the center of the zone. Again,

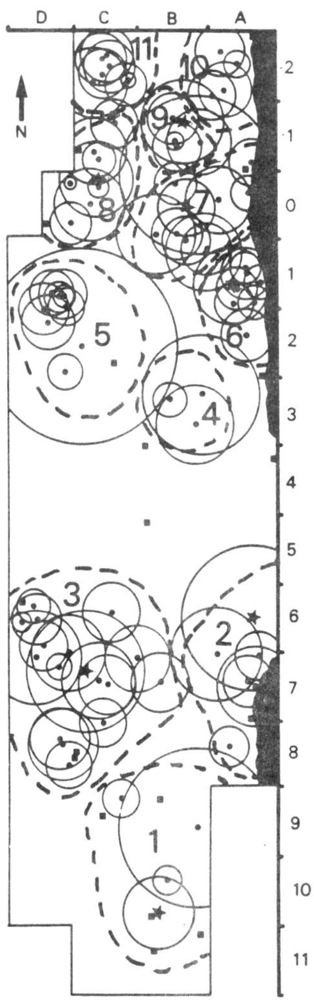

Figure 104. Eleven Zones Defined at High Scale Solution for Ten Artifact Class Distributions from Le Flageolet I Level VIII-2. Zones are outlined with a dotted line.

these cobbles are clustered just below the burnt surface of the block.

Zonation in the southern sector is essentially unchanged from the preceeding scale. The addition of more clusters merely adds to the number of centroids associated with the same three zones. Minor refinements do occur, but these amount to the reassignment of a single burin from zone 1 to zone 2 and the removal of two notched pieces from zone 3. The two solutions are identical in most respects.

More substantial changes are effected in the northern sector between the medium and high clustering scales. Two zones are added to the pattern, and zones defined at the preceeding level are more clearly delimited. A new zone is defined to the south of the A1 hearth, zone 4. This zone contains three diffuse clusters and a single isolate. There is slight overlap between zone 4 and zone 6 to the north, involving only the outer edge of zone 4's largest RMS radius. The hearth area again contains a zone composed of compact clusters (zone 6). At this scale, zone 6 is entirely separated from zones to the west, but some overlap remains with a zone centered in B0 to the north. This overlap entails only the outer edges of a few RMS circles. Perhaps because a new zone is defined to the south of zone 6, hearth area clusters are even more compact than was the case at the medium scale. Zone 5 at this scale is the concentration of clusters located in grid units D1 and D2. Again, no cobble clusters are associated with this zone. Zone 5 is made up of nine clusters here, but this increase is probably due to division between scales rather than to any addition of materials. Zone 7 at the highest scale is located just to the north of the hearth area. This is a group of six clusters and a cobble concentration scattered over a rock surface. Zone 7 overlaps somewhat with zones 6 and 9, but in each case only a few long RMS radii are involved. Away from the shelter wall and to the north of cluster zone 5, zone 8 represents a linear pattern of clusters identified at the previous scale. Zone 8 is more discrete than before but still conforms to spaces between rocks. The zone contains six clusters and an isolate, along with an isolated cobble. Zone 9, with zone 7, represents a concentration within the linear scatter of clusters north of the hearth. The zone occupies the northern end of a rock (in units B0/B-1) and contains five clusters including cobbles. Although there is RMS overlap with nearby zones, the compactness of zone 9 cluster centroids warrants their separation at this scale. The cobble cluster in zone 9 corresponds to that described in Chapter 6 as hearth materials swept from the A1 feature. Finally, two zones are retained from the previous scale. Zone 10 is located at the extreme northeast of the excavation near the wall. Four clusters and a few isolates compose the zone. There is slight overlap with zone 9 on the rock surface above but no overlap with other clusters. Zone 11 is also at the north end of excavations in the eastern corner. Six clusters are located here, none for cobbles, and there is little overlap with surrounding zones.

To summarize the results of zone definition at the highest scale, two characteristics are noted. First, cluster zones defined at the medium scale in the southern sector remain essentially unchanged. In other words, patterns in the southern distributions are defined early in the sequence. What refinements do occur between the medium and high scale maps are located in the northern sector. Three zones at the northwestern (exterior) side of the surface are discrete and confined to pockets between rocks. None of these contain cobbles in noticable quantities. Five northern zones are positioned along the shelter wall. The A1 hearth is the focus for one of these zones, and two cobble concentrations, discussed in Chapter 6 as possible hearth maintenance features, are the centers for others. Sparse zones are positioned to the north and south of these three feature-centered zones.

It should be noted that the central portion of the level VIII-2 surface is basically free of materials at all clustering scales (see Figure 104). Except for two isolated pieces in grids unit B5, none of the limited clusters that formed zones in level VIII-1 are present here. In short, the clear separation of northern from southern sectors remains throughout the clustering sequence.

Contrasts between levels VIII-1 and VIII-2 noted at the lowest clustering scale are amplified after zone definition is complete for all scales. In level VIII-1, northern sector zones are defined by the medium scale, and only the southern sector undergoes spatial refinement at the highest solution scale. The opposite holds for level VIII-2. Here, the southern zone pattern is set at the medium scale, and the northern sector sees substantial change at the high scale. The explanation of this structural difference may lie in the abundance of material present in the level VIII-2 northern sector. However, relative abundance differences between sectors were as great in level VIII-1 as in level VIII-2. There are differences between the two occupations in the apparant intensity of hearth use. The level VIII-1 feature is diffuse and shows very little evidence of maintenance or reuse. The level VIII-2 hearth is associated with a series of cobble and carbonized material accumulations that suggest more intensive use of the feature. It may be that differences between the two levels relate to the duration of occupations (resulting in hearth maintenance during the longer stay) or to different suites of activities occurring in focal areas. Only examination of accumulation contents can clarify these problems.

With results of zone definition in hand, content analyses of the level VIII-2 distributions can proceed. Particular attention will be paid to variations between the northern and southern sectors for the reasons outlined above. The nature of zones associated with the burnt block in the southern sector are also be of interest, since it may be that a second focal hearth area there creates some of the variability noted between level VIII-1 and VIII-2 spatial patterning. Given the identity in assemblage characteristics between the two occupations, variation in spatial characteristics may reflect differences in site use not evident on the level of artifact form.

STAGE III ANALYSIS: PATTERN RECOGNITION, CONTENT ANALYSIS

Because the artifact classes used in analyzing the level VIII-2 surface are the same as those employed in level VIII-1 analyses, correlation tables and content profiles are directly comparable between the two occupations. This facilitates detection of similar patterns between the layers in both global and local artifact class relations.

Assessing Global Association among Artifact Classes

Tables 39, 40, and 41 present the raw frequency counts and rank values assigned to each zone at low, medium, and high solution scales. The values are counted and transformed according to general procedures outlined in Chapter 3.

Table 42 shows the rank order correlation matrix produced for the level VIII-2 low cluster solutions. Because only two zones are considered, significance is attained only with perfect correlation. Perfect positive correlation is present among all artifact classes. This reflects the fact that all classes are most abundant in the northern sector and is a similar result to that obtained for level VIII-1.

Table 43 show rank order correlations among artifact classes at the medium scale. Three positive relationships are significant at ($p < .05$), and one negative relation is significant at that level. If a weaker level of ($p < .1$) is considered, three more relations have significant positive values. For the weaker significance level, four pairs of classes can be identified. These include: (1) truncations and cobbles, (2) endscrapers and burins, (3) utilized pieces and burins, and (4) notched pieces and debitage. Pairs (3) and (4) are distinct because endscrapers and utilized pieces have no significant relation. When ($p < .05$) defines significance, this list

TABLE 39.

Raw counts and ranks for ten artifact classes from Le Flageolet level VIII-2 based on zones defined at lowest clustering solution scale. Rank values are in parentheses.

MATERIAL CLASS	CLUSTER ZONE 1	2
Cores	9 (2)	16 (1)
Cobbles	29 (2)	184 (1)
Truncations	5 (2)	15 (1)
Burins	14 (2)	26 (1)
Endscrapers	7 (2)	27 (1)
Aurignacian Pieces	6 (2)	13 (1)
Retouched Pieces	9 (2)	37 (1)
Utilized Pieces	13 (2)	58 (1)
Notched Pieces	16 (2)	25 (1)
Debitage	282 (2)	468 (1)

TABLE 40.

Raw counts and ranks for ten artifact classes from Le Flageolet I level VIII-2 based on zones defined at the medium clustering solution scale.

MATERIAL CLASS	CLUSTER ZONE 1	2	3	4	5	6	7	8	9
					COUNTS				
Cores	2	0	7	6	4	0	6	0	0
Cobbles	5	6	18	61	0	52	14	57	0
Truncations	0	0	5	4	0	5	1	2	2
Burins	4	0	9	10	3	0	0	5	4
Endscrapers	2	0	5	8	7	0	0	12	0
Aurignacian	0	0	6	2	2	0	2	5	2
Retouched	2	0	6	0	8	30	0	0	0
Utilized	1	4	8	21	0	0	0	18	19
Notched	1	2	10	0	0	7	10	8	0
Debitage	0	23	251	105	0	141	153	77	0
					RANKS				
Cores	5	8.5	1	2.5	4	8.5	2.5	8.5	8.5
Cobbles	7	6	4	1	8.5	3	5	2	8.5
Truncations	8.5	8.5	1.5	3	8.5	1.5	6	4.5	4.5
Burins	4.5	8.5	2	1	6	8.5	8.5	3	4.5
Endscrapers	5	8.5	4	2	3	8.5	8.5	1	8.5
Aurignacian	8.5	8.5	1	3.5	3.5	8.5	3.5	2	3.5
Retouched	4	8.5	3	8.5	2	1	8.5	8.5	8.5
Utilized	6	5	4	1	8.5	8.5	8.5	3	2
Notched	6	5	1.5	8.5	8.5	4	1.5	3	8.5
Debitage	8.5	6	1	4	8.5	3	2	5	8.5

TABLE 41.

Raw counts and ranks for ten artifact classes from Le Flageolet I level VIII-2 based on zones defined at the highest clustering solution scale.

MATERIAL CLASS	CLUSTER ZONE 1	2	3	4	5	6
			COUNTS			
Cores	2	4	3	2	2	3
Cobbles	5	6	18	0	0	61
Truncations	0	0	5	1	2	6
Burins	3	1	9	4	3	7
Endscrapers	2	0	5	0	7	7
Aurignacian	0	0	6	0	2	2
Retouched	2	0	6	0	8	0
Utilized	1	4	8	2	2	0
Notched	1	2	12	0	4	0
Debitage	0	23	251	0	0	105
			RANKS			
Cores	4.5	1	2.5	4.5	4.5	3.5
Cobbles	7	6	4	10.5	10.5	1
Truncations	10.5	10.5	2	6.5	4.5	1
Burins	6.5	8	1	4.5	6.5	2
Endscrapers	6.5	10.5	3.5	10.5	1.5	1.5
Aurignacian	10.5	10.5	1	10.5	3.5	3.5
Retouched	4	10.5	3	10.5	2	10.5
Utilized	9	6	4	7.5	7.5	10.5
Notched	8	6.5	3	10.5	4.5	10.5
Debitage	10.5	6	1	10.5	10.5	4

TABLE 41 (Continued).

MATERIAL CLASS	CLUSTER ZONE 7	8	9	10	11
			COUNTS		
Cores	3	3	0	0	2
Cobbles	52	14	57	0	
Truncations	3	1	2	0	2
Burins	6	0	0	2	4
Endscrapers	2	2	5	0	4
Aurignacian	1	2	0	4	2
Retouched	30	0	0	0	0
Utilized	18	5	16	0	15
Notched	6	6	4	2	2
Debitage	141	153	0	77	0
			RANKS		
Cores	2.5	2.5	10.5	10.5	4.5
Cobbles	3	5	2	10.5	10.5
Truncations	3	6.5	4.5	10.5	4.5
Burins	3	10.5	10.5	7	4.5
Endscrapers	6.5	6.5	3.5	10.5	5
Aurignacian	5	3.5	10.5	2	3.5
Retouched	1	10.5	10.5	10.5	10.5
Utilized	1	5	2	10.5	3
Notched	2.5	2.5	4.5	6.5	6.5
Debitage	3	2	10.5	5	10.5

TABLE 42.

Matrix of Spearman's rank order correlation statistics (rho) for all Le Flageolet I level VIII-2 artifact classes when two cluster zones are considered. Artifact classes are numbered as follows: (1) cores, (2) cobbles, (3) truncations, (4) burins, (5) endscrapers, (6) Aurignacian pieces, (7) miscellaneous retouched pieces, (8) utilized pieces, (9) notched pieces, and (10) debitage. All correlations of 1.0 have ($p < .01$).

	1	2	3	4	5	6	7	8	9	10
1	/									
2	1.0	/								
3	1.0	1.0	/							
4	1.0	1.0	1.0	/						
5	1.0	1.0	1.0	1.0	/					
6	1.0	1.0	1.0	1.0	1.0	/				
7	1.0	1.0	1.0	1.0	1.0	1.0	/			
8	1.0	1.0	1.0	1.0	1.0	1.0	1.0	/		
9	1.0	1.0	1.0	1.0	1.0	1.0	1.0	1.0	/	
10	1.0	1.0	1.0	1.0	1.0	1.0	1.0	1.0	1.0	/

TABLE 43.

Matrix of Spearman's rank order correlation statistics (rho) for all Le Flageolet I level VIII-2 artifact classes when nine cluster zones are considered. Artifact classes are numbered as follows: (1) cores, (2) cobbles, (3) truncations, (4) burins, (5) endscrapers, (6) Aurignacian pieces, (7) miscellaneous retouched pieces, (8) utilized pieces, (9) notched pieces, and (10) debitage. Rho values with (p < .05) are starred.

	1	2	3	4	5	6	7	8	9
1	/								
2	-.00	/							
3	-.05	.63	/						
4	.25	.31	.30	/					
5	.15	.29	-.11	.71*	/				
6	.32	.11	.27	.53	.37	/			
7	-.14	-.44	-.11	-.47	-.32	-.66	/		
8	-.28	.27	.23	.74*	.24	.28		- .76*	/
9	-.01	.34	.22	-.35	-.38	-.01	-.24	-.45	/
10	.30	.68	.65	-.09	-.25	.18	-.26	-.20	.73*

is reduced to three pairs: (1) endscrapers and burins, (2) utilized pieces and burins, and (3) notched pieces and debitage. Three artifact classes (cores, Aurignacian pieces, and miscellaneous retouched pieces) have no positive associations with other classes. Retouched and utilized pieces have a high negative correlation at the medium scale.

Correlation between burins and endscrapers seems to be determined by their coincidence in zones at and to the south of the hearth. Both have high ranks in zone 4 centered on the feature. Similar rank values characterize both classes in southern sector zones. In the north, zone 8 has similar ranks for the two classes, but there does not seem to be regular coincidence in other northern sector accumulations. Thus, similar frequencies of burins and endscrapers in the southern sector seem to contribute to their global association at the medium scale.

Correlation between burins and utilized pieces must occur in other zones (given the lack of association between endscrapers and utilized pieces). In the south, rank values are close in zones 1 and 3, although not as close as for burins and endscrapers. However, rank values are similar throughout the northern sector. The highest ranks for both classes occur in the hearth area zone 4. Thus, the pattern that emerges has burins distributed over the entire surface and associated with endscrapers in the south and utilized pieces in the north.

Debitage and notched pieces have their highest rank values in southern zone 3. This contrasts with the other correlated pairs, which all have their highest rank values in the northern sector. Artifact classes not correlated with other categories also have their highest ranking clusters in southern zone 3, including cores, truncations, and Aurignacian pieces. Other high ranking zones for debitage and notched pieces occur in the northern sector, and their ranks are low in other southern sector zones. It is interesting to note that, at least at the medium scale, notched pieces are absent from the hearth area zone 4. Debitage, on the other hand, has a middle range rank value for this zone. The hearth zone provides one of the only loci of rank departure between notched pieces and debitage, reducing the overall correlation value for the pair.

Rank order correlations at the medium scale indicate some differences in spatial patterning among level VIII-2 artifact classes. Some classes have their highest ranks in the hearth zone, while others have their highest ranks in the southern sector of the surface. Retouched pieces and endscrapers diverge from this in that their highest frequencies occur in northern sector zones away from the hearth. In a more general sense, it should be noted that the number of significant positive correlations drops considerably between the low and medium scales of refinement. Only three positive values have ($p < .05$) here. At the low scale, all relations are positive and significant.

Table 44 presents the correlation matrix for level VIII-2 divided into eleven zones. Only two artifact class pairs are indicated here, and these correlations are significant at ($p < .05$). The first pair comprises endscrapers and truncations. Both of these have high rank values in southern zone 3, hearth zone 6, and zone 9 to the north of the hearth area. A second correlated set includes notched and utilized pieces. These have high ranks in southern zone 3 and in zones 7 and 9 north of the hearth against the shelter wall.

High rank values for artifact classes having no significant correlations at this scale occur primarily in three zones. All artifact classes are abundant in zone 3. Seven classes have ranks between 1 and 4 in hearth zone 6. Nine classes have these ranks just to the north of the hearth in zone 7. It is interesting to note that the group of artifact classes labelled "expedient" (notched, retouched, and utilized pieces) are absent from the hearth zone proper. These classes occur in large numbers in zone 7 just north of the feature. More labor-intensive forms have high ranks <u>at</u> the hearth and lower ranks away from it. If, as was suggested earlier, the materials located on the block surface to the north of the hearth represent hearth sweeping accumulations, then it would seem plausible to suggest that expedient tools were used while the hearth was in operation and were therefore part of the assemblage swept from it. Labor intensive items were used either at the feature late in the occupation or were deposited into the hearth at abandonment.

Assuming that the northern features represent a focal area "complex", the composition of zones 6, 7, and 9 may help to explain southern zone 3. Zone 3 contains large numbers of all tool forms. If materials generated at a hearth on the rocks above were deposited into the cracks below, by either human or natural agents, then it might be that hearth maintenance accumulations and final hearth contents were mixed. Such an accumulation might be very much like zone 3 and its contents similar to the northern feature complex as a whole. If such similarity can be demonstrated, the notion that there was a second level VIII-2 focal hearth in the southern sector would be supported. We shall return to this matter later.

There are apparant differences between the northern and southern sectors in patterns of interclass correlation. These diverge from the marked variation between sectors noted in level VIII-1. For example, burins are more evenly distributed and have correlations with other classes in both sectors of the surface. Endscrapers are also associated with other classes here, which was not the case for level VIII-1. How do the two

TABLE 44.

Matrix of Spearman's rank order correlation statistics (*rho*) for all Le Flageolet I level VIII-2 artifact classes when eleven cluster zones are considered. Artifact classes are numbered as follows: (1) cores, (2) cobbles, (3) truncations, (4) burins, (5) endscrapers, (6) Aurignacian pieces, (7) miscellaneous retouched pieces, (8) utilized pieces, (9) notched pieces, and (10) debitage. *Rho* values with (p < .05) are starred.

	1	2	3	4	5	6	7	8	9
1	/								
2	-.01	/							
3	.16	.39	/						
4	.24	.08	.39	/					
5	-.01	.48	.74*	.00	/				
6	-.12	-.35	.33	.12	.25	/			
7	-.36	-.36	-.13	-.12	.03	-.29	/		
8	-.03	.22	.36	-.06	.13	-.22	.11	/	
9	-.09	.18	.25	-.16	.25	.34	.20	.68*	/
10	.12	.36	.07	.19	-.24	.39	-.31	-.11	.37

Couche VIII levels compare in terms of homogeneity?

Figure 105 depicts the homogeneity plot for level VIII-2 based on the global correlations defined at three refinement scales. As can be seen, this plot is very similar to that obtained for level VIII-1 and different from the Pincevent graph. At low scales, all artifact classes are intercorrelated. As finer spatial units are examined, fewer correlations are present. It has been argued that this pattern reflects the isolation of specialized activity areas at fine scales of spatial resolution. It seems that, in level VIII-2, variability in non-hearth area accumulations defined at high scales of resolution reduces the homogeneity of zone contents.

Variation in level VIII-2 zone content is complex. Complexity may or may not relate to differences between this and the level VIII-1 occupation. There are reasons to believe both that the two occupations are different (e.g., different patterns of cluster formation in the two site sectors) and that they are similar (e.g., changing homogeneity profiles). The resolution of this problem may be addressed by examining local associations among artifact classes. What is the relation between the northern hearth area deposits and the southern sector accumulations in level VIII-2?

Assessing Local Association among Artifact Classes

Local variability in zone content is examined using percent of maximum standardizations as in previous analyses. Table 45 presents class maximum standardizations for level VIII-2 when two zones are considered. Figure 106 illustrates these values with circular graphs. As is expected, zone 2 in the north has maximum values for all classes, and its graphic profile has the circular form resulting from such a pattern. Southern zone 1 has relatively high values for five classes, and these show as peaks on Figure 106. Debitage, notched pieces, burins, cores, and Aurignacian pieces are all well represented in zone 1. Class maximum values suggest that differences between the northern and southern sectors are primarily in overall material abundance. The northern sector is considerably richer in all materials than is the south. Cobbles, moreover, seem to be very concentrated in the northern sector. If there was a hearth located in the south, this characteristic would suggest that it was of less importance than the northern feature.

Table 46 shows zone maximum standardization values for the level VIII-2 low scale zones. Figure 107 illustrates these data. (In zone maximum tables throughout level VIII-2 content analyses, debitage and cobbles will both be eliminated from calculations. This

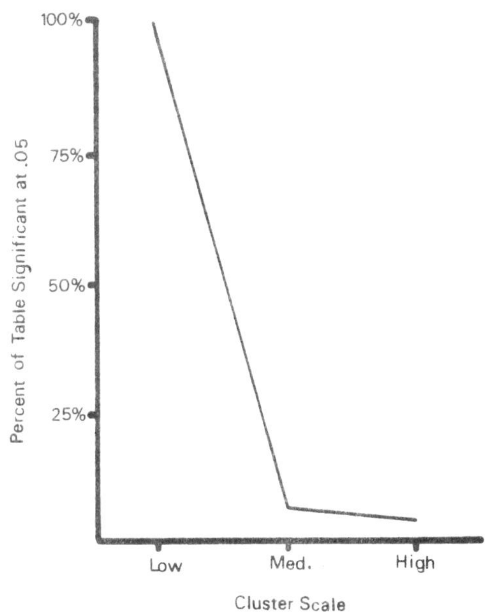

Figure 105. Plot of Accumulation Content Homogeneity with Scale of Spatial Refinement for Le Flageolet I Level VIII-2.

TABLE 45.

Matrix of class maximum standardization values for Le Flageolet I level VIII-2 artifact classes when two cluster zones are considered. All values are rounded to the nearest integer.

MATERIAL CLASS	CLUSTER ZONE 1	2
Cores	56	100
Cobbles	16	100
Truncations	33	100
Burins	54	100
Endscrapers	26	100
Aurignacian	46	100
Retouched	24	100
Utilized	22	100
Notched	64	100
Debitage	60	100

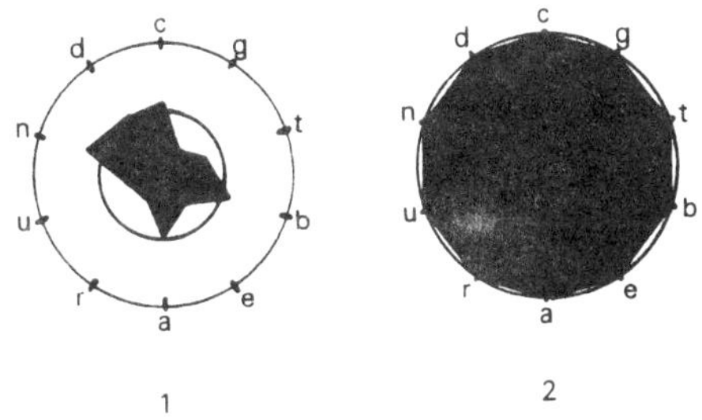

Figure 106. Circular Graphs for Class Maximum Standardization of Zone Contents when Two Zones Are Considered for Level VIII-2. Radii correspond to artifact classes as follows: (c) cores, (g) cobbles, (t) truncations, (b) burins, (e) endscrapers, (a) Aurignacian pieces, (r) miscellaneous retouched pieces, (u) utilized pieces, (n) notched pieces, (d) debitage. Circumference of circle represents class maximum value of 100.

TABLE 46.

Matrix of zone maximum standardization values for Le Flageolet I level VIII-2 artifact classes when two cluster zones are considered. All values are rounded to the nearest integer. Note that debitage and cobbles have been excluded from calculations.

MATERIAL CLASS	CLUSTER ZONE 1	CLUSTER ZONE 2
Cores	56	28
Truncations	31	26
Burins	88	45
Endscrapers	44	47
Aurignacian	38	22
Retouched	56	64
Utilized	81	100
Notched	100	43

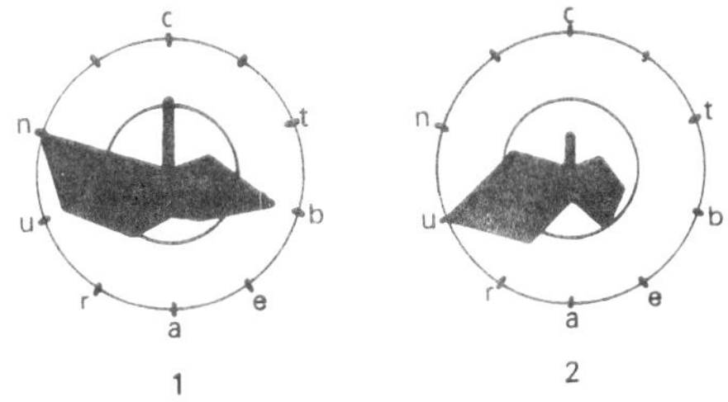

Figure 107. Circular Graphs for Zone Maximum Standardization of Zone Contents when Two Zones Are Considered for Level VIII-2. Radii correspond to artifact classes as follows: (c) cores, (t) truncations, (b) burins, (e) endscrapers, (a) Aurignacian pieces, (r) miscellaneous retouched pieces, (u) utilized pieces, (n) notched pieces. Circumference of circle represents class maximum value of 100.

is because they are so abundant that they overwhelm other classes in every zone maximum profile). The two profiles are similar between northern and southern zones, but there are a few artifact classes that seem to concentrate in one sector or the other. Burins, cores, and notched pieces show peaks in southern zone 1, but the first two classes also show peaks of lesser magnitude in zone 2. In most respects, zone maximum standardization suggests that the northern and southern zones in level VIII-2 are similar in content. Both class maximum values and correlation analysis support this view and indicate that differences among zones at the low scale are in material abundance rather than in content of accumulations.

Percent of maximum standardizations at the medium scale for level VIII-2 show more complex relations among classes. Table 47 gives class maximum values for the nine zone configuration, and these are displayed graphically in Figure 108. Three zones have high values for many of the artifact classes considered. Zone 3 in the southern sector dominates the list, although the profiles for utilized and miscellaneous retouched pieces indicate a reduced presence for these classes. Cores, truncations, burins, Aurignacian pieces, and notched and denticulated pieces have peaks in zone 3. The hearth area zone 4 is also rich, with high values for most classes but relatively few expedient tools. Zones 6 and 8 to the north of the hearth, tentatively interpreted as concentrations within a single hearth complex, contain quantities of a number of classes. Zone 6 has only five classes present, but four of these (truncations, cobbles, miscellaneous retouched pieces, and notched pieces) are abundant. Zone eight has high standardized values for five of the eight classes present: cobbles, endscrapers, Aurignacian pieces, utilized pieces, and notched pieces. The truncations and retouched pieces having peaks in zone 6 are infrequent in zone 8. On the other hand, endscrapers, utilized pieces, and Aurignacian pieces are abundant in zone 8 in contrast with the zone 6 assemblage. Zone 7 is different from the rich zones in that it contains middle-range values for a few classes. However, zone 7 shows peaks for cores, debitage, and notches. This zone also contains quantities of dechet (see Chapter 6) and may represent (in part) a knapping waste disposal zone.

Other zones must be considered sparse based on class maximum values. Zones 1 and 2 (in the south) and 5 and 9 (in the north) contain a varied and limited set of artifact classes. Zones 1 and 2 share low values for utilized and notched pieces, as well as for cobbles. Zone 1 also contains a few cores, burins, and endscrapers. Northern sector zone 5 combines cores and endscrapers. Zone 9 has a very high value for utilized pieces and low values for three other classes.

TABLE 47.

Matrix of class maximum standardization values for Le Flageolet I level VIII-2 artifact classes when nine cluster zones are considered. All values are rounded to the nearest integer.

MATERIAL CLASS	CLUSTER ZONE 1	2	3	4	5	6	7	8	9
Cores	29	0	100	86	57	0	86	0	0
Cobbles	8	10	30	100	0	85	23	93	0
Truncations	0	0	100	80	0	100	20	40	40
Burins	40	0	90	100	30	0	0	50	40
Endscrapers	17	0	42	67	58	0	0	100	0
Aurignacian	0	0	100	33	33	0	33	83	33
Retouched	7	0	20	0	27	100	0	0	0
Utilized	5	19	38	100	0	0	0	86	90
Notched	10	20	100	0	0	70	100	80	0
Debitage	0	9	100	42	0	56	61	31	0

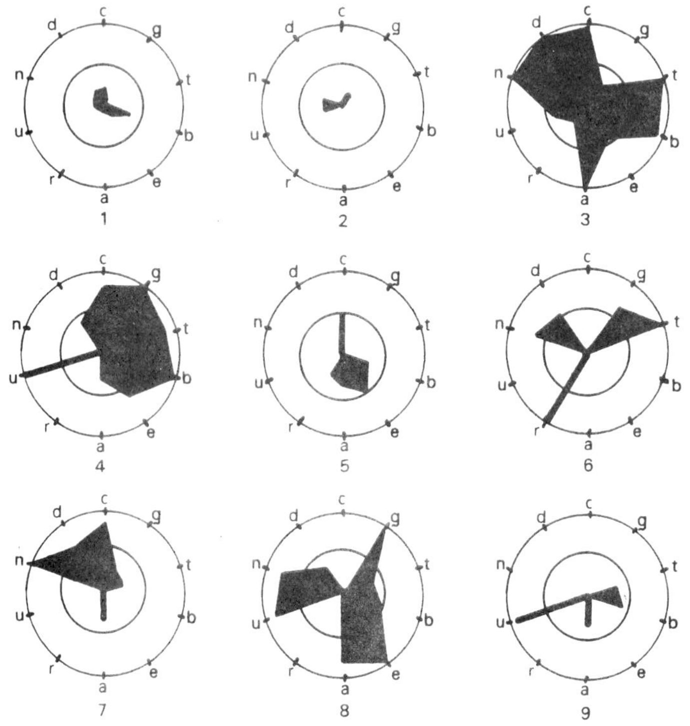

Figure 108. Circular Graphs for Class Maximum Standardization of Zone Contents when Nine Zones Are Considered for Level VIII-2. Radii correspond to artifact classes as follows: (c) cores, (g) cobbles, (t) truncations, (b) burins, (e) endscrapers, (a) Aurignacian pieces, (r) miscellaneous retouched pieces, (u) utilized pieces, (n) notched pieces, (d) debitage. Circumference of circle represents class maximum value of 100.

Table 48 lists zone maximum percentages for the level VIII-2 medium solution. Figure 109 displays these data. A great deal of variation in zone content is evident. The profile from zone 3 is similar to that obtained at the low clustering scale for the southern sector. A general assemblage is present dominated by notches, burins, and utilized pieces. Zone 4 lacks notched and miscellaneous retouched pieces and has a peak for utilized pieces. In other respects zone 4 (the hearth) is similar to zone 3. Zone 4 also has a profile similar to zones 8 and 9. All three are dominated by utilized pieces and have few truncations, burins, or Aurignacian pieces. Zone 4 contains cores that the other two lack; zone 8 has notched pieces absent in 4 and 9; and zone 9 lacks the endscrapers that are a component of the other two assemblages. While zone 6 was similar to some of these based on class maximum percentage profiles, it is distinct here because miscellaneous retouched pieces are dominant.

Although lacking Aurignacian pieces and truncations, sparse zone 1 is similar to neighboring zone 3 in other respects. However, the assemblage for zone 1 is so small that stochastic processes discussed in Chapter 3 may be in effect, i.e., the zone profile would be altered with the addition of a single item. Zone 2 in the southern sector contains a few utilized and notched pieces in similar proportion to zone 8. However, other elements present in zone 8 are absent here, and the assemblage size is, again, very small.

Finally, zones 5 and 7 appear to be different from other zones in content. Zone 5 has peaks for utilized and Aurignacian pieces. Zone 7 is dominated by notched pieces and cores. These zones are tentatively interpreted as dump accumulations associated with the northern hearth complex. Zone 5 is similar to the hearth zone 4 in its profile for the two classes just listed.

In summarizing the results of content analysis at the medium clustering scale, two axes of variation can again be invoked. Material abundance can be used to define three varieties of zone. Rich zones include numbers 3, 4, 6, and 8. Zone 4 is centered on the northern sector hearth, and zones 6 and 8 are interpreted as hearth maintenance features associated with zone 4. Southern sector zone 3 is a very general concentration that may have resulted from deposition of both hearth maintenance and feature contents into a single accumulation. If this is so, the feature was probably located on the rocks above. Medium-rich zones include 5 and 7 (both in the northern sector). These accumulations are located along the eastern part of the northern sector and are interpreted as hearth area dumps. Finally, a group of sparse zones (1, 2, and 9) contain few objects from a limited number of classes.

TABLE 48.

Matrix of zone maximum standardization values for Le Flageolet I level VIII-2 artifact classes when nine cluster zones are considered. All values are rounded to the nearest integer. Note that cobbles and debitage have been excluded from calculations.

MATERIAL CLASS	CLUSTER ZONE 1	2	3	4	5	6	7	8	9
Cores	50	0	70	29	50	0	60	0	0
Truncations	0	0	50	19	0	17	10	11	11
Burins	100	0	90	48	38	0	0	28	21
Endscrapers	50	0	50	38	88	0	0	68	0
Aurignacian	0	0	60	10	25	0	20	28	11
Retouched	50	0	60	0	100	100	0	0	0
Utilized	25	100	80	100	0	0	0	100	100
Notched	25	50	100	0	0	23	100	44	0

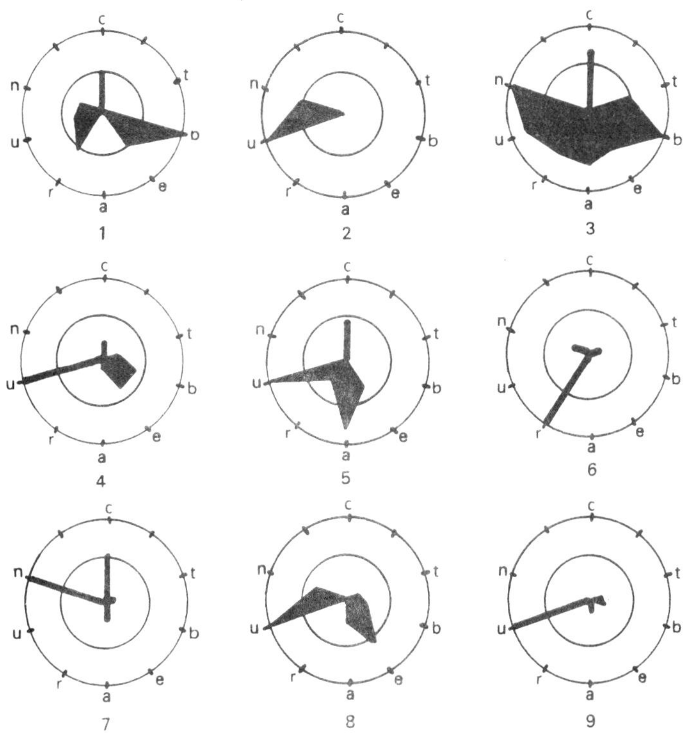

Figure 109. Circular Graphs for Zone Maximum Standardization of Zone Contents when Nine Zones Are Considered for Level VIII-2. Radii correspond to artifact classes as follows: (c) cores, (t) truncations, (b) burins, (e) endscrapers, (a) Aurignacian pieces, (r) miscellaneous retouched pieces, (u) utilized pieces, (n) notched pieces. Circumference of circle represents class maximum value of 100.

A second axis of variation concerns local assemblage content. Zones 4 and 8 in the northern sector are similar, but this is likely if they constitute parts of a single hearth complex. More surprising, perhaps, is the similarity in zone maximim profiles between zone 3 (a rich zone) and sparse zone 1. Southern sector zones show similar content profiles in level VIII-1. A comparison of the VIII-1 southern zones to the level VIII-2 zones shows that the southern accumulations are also similar between occupations (compare Figures 88 in Chapter 7 to Figure 109 here). There are differences in spatial pattern between the two levels, however. In level VIII-1, southern zones form a linear configuration around the central block. In level VIII-2, a single zone is present at the northern end of the same block. It may be that zone 3 at the level VIII-2 medium scale does not represent a second focal hearth but a more compact version of the level VIII-1 southern zones. Zone maximum profiles would support such an interpretation. Conversely, the presence of cobbles in VIII-2 zone 3 (admittedly in low frequency when compared to the northern zones) and a different pattern of small artifact accumulation, suggests that it may represent a hearth deposit.

One way to examine this interpretive problem is by comparing zone 3 with more certain hearth accumulations. It has been suggested that zone 3 may represent a redeposited mixture of hearth contents and sweepings. Zone 4 is the only certain hearth area zone in level VIII-2, and its contents are more restricted that those for zone 3. Zones 6 and 8 have been interpreted as deposits generated by cleaning out the Al hearth. By combining class frequencies for zones 4, 6, and 8, a more general profile for a "hearth complex" can be prepared for comparison with zone 3. Differences between two profiles, one a combination and the other zone 3, would reduce the plausibility of inferring a focal hearth location in the southern sector.

Figure 110 shows the profiles generated for comparison after zone maximum standardization. The profile for zone 3 is the same as presented in Figure 109. These graphs are certainly more similar than are the individual northern sector zone profiles compared to that for zone 3. However, important differences exist that preclude interpretation of zone 3 as a hearth deposit. First, zone 3 has quantities of burins, cores, and Aurignacian pieces that are greater than those in the hearth area accumulations. These are the same tools found in abundance in the level VIII-1 southern sector zones. Second, the combination of zones 4, 6, and 8 here produce a profile very similar to the single hearth accumulation defined for level VIII-1 (see Figure 93, zone 8). While this supports the interpretation of these northern zones as parts of a hearth complex, it does not support the inference of a second feature near zone 3. It seems more likely that zone 3 represents an overlapping concentration

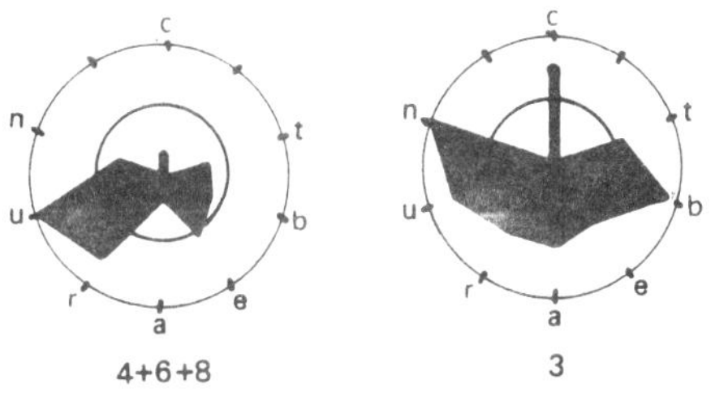

Figure 110. Circular Graphs for Zone Maximum Standardization of Zone Contents when Level VIII-2 Medium Scale Zones 4, 6, 8, Are Combined and Compared to Zonc 3. Radii correspond to artifact classes as follows: classes as follows: (c) cores, (g) cobbles, (t) truncations, (b) burins, (e) endscrapers, (a) Aurignacian pieces, (r) miscellaneous retouched pieces, (u) utilized pieces, (n) notched pieces, (d) debitage. Circumference of circle represents class maximum value of 100.

of deposits similar to those having a linear configuration in the VIII-1 southern sector. The content of southern sector zones is similar in both levels, but their spatial pattern is different. Hearths, on the other hand, also share assemblage characteristics between the two occupations and are distinct from the southern sector zones in similar ways.

High scale solutions for level VIII-2 refine this picture of zone content, but the basic characteristics are maintained. Table 49 shows class maximum standardization values for eleven cluster zones, and Figure 111 illustrates these graphically. Only three zones can be identified as rich at this scale: zones 3, 6, and 7. The first is a replicate of zone 3 at the preceeding scale and needs no further elaboration. Zone 6 centers on the A1 hearth and has abundant cores, truncations, burins, endscrapers, Aurignacian pieces, notched pieces, and debitage. Except for the last two classes named, these are all labor-intensive tool forms. Zone 7 also contains cores, and here these are associated with cobbles, utilized pieces, and miscellaneous retouched pieces. The last two classes represent expediently produced items. As has been discussed, the zone 3 assemblage is dominated by labor-intensive tools.

Four zones are medium-rich at this scale, including zones 5, 8, 9, and 11. All are located in the northern sector of the site. Three of these were tentatively identified as dump zones at the previous solution scale (5, 8, and 11). The fourth is located at the north end of the large block underlying the hearth complex. These zones vary in content. Zone 5 shows a peak for endscrapers. Zone 8 has a high value for cores. Utilized pieces dominate the zone 11 class maximum profile. A more complex assortment of endscrapers, utilized pieces, and cobbles composes zone 9. These zones show low values for most other artifact classes.

A third group of sparse zones is again defined. These uniformly show low values for most classes and are typically dominated by a few classes. Zones 1 and 2 in the southern sector fall into this group and have similar profiles dominated by cores, notches, utilized pieces, and burins. In the northern sector, zones 4 and 10 are sparse. Zone four contains some burins and cores along with a few truncations and utilized pieces. Zone 10 holds a few Aurignacian pieces, burins, notches, and debitage.

The spatial distribution of these class maximum zone categories parallels that noted in level VIII-1. Sparse zones are located at the periphery of the occupation surface. Rich zones are located at, north, and west of the focal hearth. Medium-rich zones are located near but away from the hearth zone in pockets between rocks. A difference in spatial pattern between levels is present in the southern sector where a single rich zone replaces the

TABLE 49.

Matrix of class maximum standardization values for Le Flageolet I level VIII-2 artifact classes when eleven cluster zones are considered. All values are rounded to the nearest integer.

ARTIFACT CLASS	CLUSTER ZONE 1	2	3	4	5	6	7	8	9	10	11
Cores	50	100	75	50	50	75	75	75	0	0	50
Cobbles	8	10	30	0	0	100	85	23	93	0	0
Truncations	0	0	83	17	33	100	50	17	33	0	33
Burins	33	11	100	44	33	78	67	0	0	22	44
Endscrapers	29	0	71	0	100	100	29	29	71	0	57
Aurignacian	0	0	100	0	33	33	17	33	0	67	33
Retouched	7	0	20	0	27	0	100	0	0	0	0
Utilized	6	22	44	11	11	0	100	28	89	0	83
Notched	8	17	100	0	33	0	50	50	33	17	17
Debitage	0	9	100	0	0	42	56	61	0	31	0

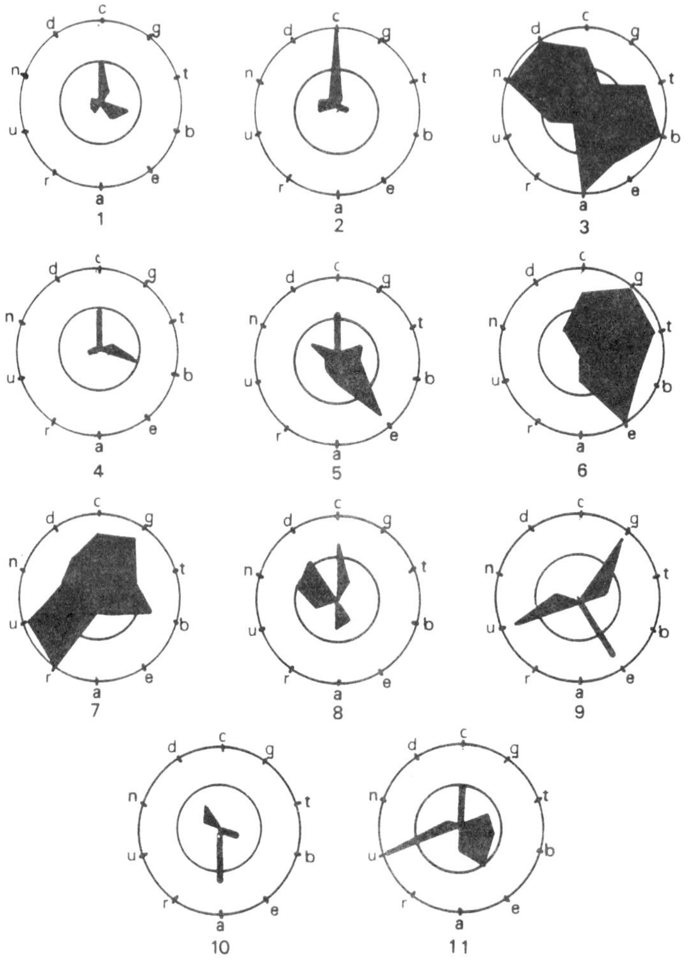

Figure 111. Circular Graphs for Class Maximum Standardization of Zone Contents when Eleven Zones Are Considered for Level VIII-2. Radii correspond to artifact classes as follows: (c) cores, (g) cobbles, (t) truncations, (b) burins, (e) endscrapers, (a) Aurignacian pieces, (r) miscellaneous retouched pieces, (u) utilized pieces, (n) notched pieces, (d) debitage. Circumference of circle represents class maximum value of 100.

several sparse zones present at this scale in level VIII-1. This may represent the repeated deposition of refuse into a single area in level VIII-2 while refuse was scattered in VIII-1. Southern zone content would support such an interpretation.

Zone maximum standardizations for the level VIII-2 highest scale solution are presented in Table 50 and Figure 112. Zones 1, 2, and 3 in the south are unchanged from the previous scale and need no detailed discussion. They are three similar deposits dominated by burins and cores and different in content from hearth area zones. Stochastic processes may effect profiles for sparse zones 1 and 2.

Rich zones in the northern sector have the same profile characteristics as those described for medium scale zones. The addition of two new zones in the sector does change some of the profiles. Zone 6 at the hearth lacks the utilized pieces that were present at the lower scale of resolution. These objects have been incorporated into zone 7 to the north. A decrease in utilized pieces and the addition of burins, miscellaneous retouched pieces, truncations, and notches characterizes zone 5 to the west of the A1 hearth. These changes are probably due to the formation of zone 4 and the refinement of zone boundaries between 5, 6, and 8 at this scale. The zone designated number 8 at the medium scale is now divided into two zones. Zone 9 (occupying roughly the same place as the previous zone 8) retains a profile similar to its predecessor. Zone 10 to the northeast is a sparse accumulation containing a few notches, burins, and Aurignacian pieces. Two rich zones associated with the hearth retain their zone maximum profiles from the medium solution scale. Some changes are apparant; utilized pieces are added to the zone just north of the hearth (here zone 7) and to the one west of that area (zone 8).

The overall picture of variation in zone content that emerges from the level VIII-2 high scale solutions does not differ much from medium scale results. Southern sector zonation does not change between refinement scales. Northern sector patterns are refined around the A1 hearth. At and to the south of the feature, labor intensive tools, along with cores, dominate local assemblages. To the west of the focal feature (in zone 5) endscrapers and retouched pieces are concentrated. The profile for this zone is similar to that for the C0/D0 dump in level VIII-1. A similar interpretation is offered for zone 5 here. Zones to the north of the hearth contain quantities of expedient tool forms. These are concentrated at and near the hearth in level VIII-1. Overall, variation in spatial organization and zone content is very similar between the two Couche VIII occupation surfaces.

TABLE 50.

Matrix of zone maximum standardization values for Le Flageolet I level VIII-2 artifact classes when eleven cluster zones are considered. All values are rounded to the nearest integer. Note that debitage and cobbles have been excluded from calculations.

ARTIFACT CLASS	CLUSTER ZONE 1	2	3	4	5	6	7	8	9	10	11
Cores	67	100	25	50	25	43	10	50	0	0	13
Truncations	0	0	42	25	25	86	10	17	13	0	13
Burins	100	25	75	100	38	100	20	0	0	50	27
Endscrapers	67	0	42	0	88	100	7	33	31	0	27
Aurignacian	0	0	50	0	25	29	3	33	0	100	13
Retouched	67	0	50	0	100	0	100	0	0	0	0
Utilized	33	100	67	50	25	0	60	83	100	0	100
Notched	33	50	100	0	50	0	20	100	25	50	13

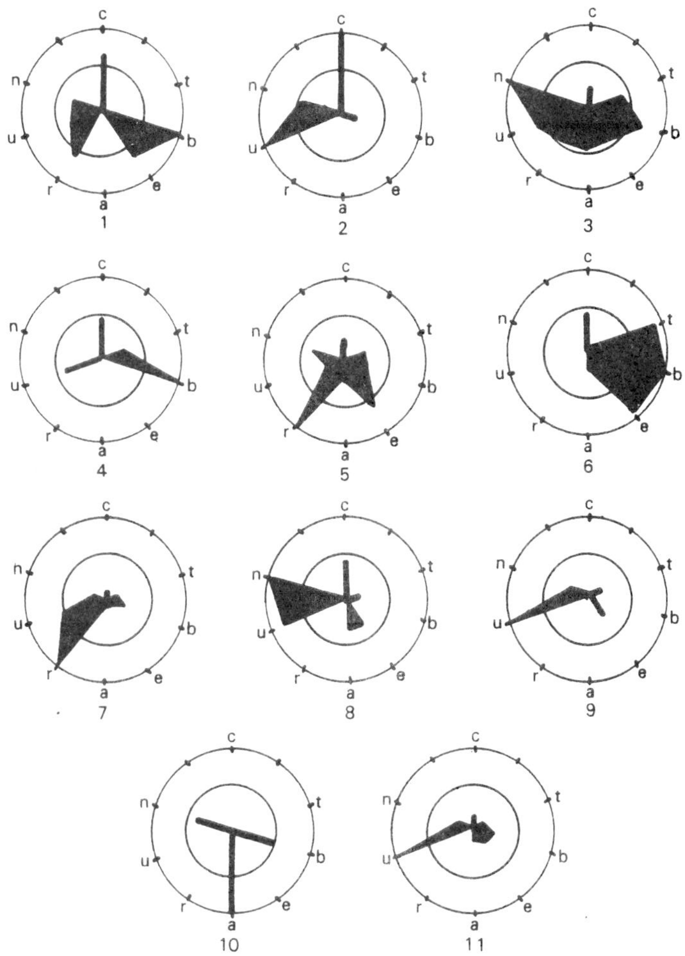

Figure 112. Circular Graphs for Zone Maximum Standardization of Zone Contents when Eleven Zones Are Considered for Level VIII-2. Radii correspond to artifact classes as follows: (c) cores, (t) truncations, (b) burins, (e) endscrapers, (a) Aurignacian pieces, (r) miscellaneous retouched pieces, (u) utilized pieces, (n) notched pieces. Circumference of circle represents class maximum value of 100.

As was discussed earlier, one of the most interesting differences between the two occupations concerns the sequence of zone formation in northern and southern sectors of the site. In level VIII-1, northern sector zonation is essentially complete by the medium solution scale, and the southern sector pattern is refined between medium and high scales. In level VIII-2, the opposite condition obtains; zones in the southern sector are defined at the medium scale, and northern patterns undergo change between medium and high scale solutions. This difference may reflect variation in the mode of deposition in the southern sectors of each occupation. Southern sector zones in level VIII-1 are similar in content and represent a series of sparse accumulations organized in a linear pattern around the base of a large block. A single, rich zone is present in the level VIII-2 southern sector (along with two sparse zones) located at the northern end of the block. However, the content of this rich zone is similar to other level VIII-2 southern sector zones and to level VIII-1 southern accumulations. This suggests that the discrete but similar deposits defined in level VIII-1 are simply layered into a single accumulation in level VIII-2. In short, the two zones are functionally similar, but refuse was disposed of all around the block in VIII-1 and off one end in VIII-2.

Zone organization and content in the northern sector is very similar between the two levels. However, hearth area deposits are more complex in level VIII-2 than in VIII-1. The intensity of hearth maintenance seems to have been greater in level VIII-2, resulting in a series of sweepings accumulations not encountered in the later occupation. Another indication of more intensive hearth area use is the series of dump zones paralleling the hearth complex. These are richer than dump zones defined in level VIII-1 but are similar in content. The spatial pattern of accumulations around the focal hearth is also similar between the two occupation surfaces. Hearths and hearth-related materials are located near the shelter wall, and hearth-area dumps are positioned in cracks between rocks towards the shelter exterior.

DISCUSSION AND CONCLUSIONS

As is the case for level VIII-1, a general dichotomy between deposits in the north and south sectors of the VIII-2 surface is indicated. Southern sector accumulations lack fine scale patterns in level VIII-2, and this may relate to a different refuse disposal mode from level VIII-1. Assemblage comparisons between southern zones from the two levels reveal similar content profiles. High proportions of burins, cores, and utilized pieces characterize the southern sector in both occupations. Although the richness and diversity of materials present in level VIII-2 zone 3 led to initial interpretations of the deposit as a second focal hearth area, content analysis does not support such an inference. Rather, the southern sector of level VIII-2 seems to reflect a limited range of activities similar to the level VIII-1 southern accumulations. It is possible that a small hearth (responsible for the accumulation of cobbles and small rock fragments in unit C7) was located on a rock surface in the south and that this hearth served as a focus for the special acitvities. If such a focus was present, it could account for the deposition of refuse in a single pile near the feature rather than in a dispersed series of small-scale dumps. In any case, level VIII-2 southern sector deposits seem to represent a pamlimpsest of refuse generated in specific activities.

In the level VIII-2 northern sector, accumulations are similar to that sector in level VIII-1. An intensively-used focal hearth complex is indicated by multiple concentrations of hearth maintenance refuse. The organization of the hearth area and associated dumps is identical in the two occupations. Artifact classes deposited directly into the hearth are dominated by labor intensive forms. Expedient tools are most common in surrounding accumulations, especially in hearth maintenance deposits. As was the case in level VIII-1, endscrapers seem to have been disposed of in dumps more often than into the hearth itself even though they are labor intensive tools. This pattern suggests that expedient tools were used at the hearth while it was in operation, and "dead" (no longer useful) intensive tools were discarded into the feature upon site abandonment during both occupations.

Figure 113 gives an interpretive map of the level VIII-2 surface. The map is very similar to that offered for level VIII-1. Table 51 lists the within class percentages for each artifact class deposited into each inferred deposit type. A comparison between this table and Table 36 (Chapter 7) emphasizes the similarity between Couche VIII occupations. However, a few differences are present. While expedient tools were associated with the hearth itself in VIII-1, these are most frequent in "sweepings" accumulations in VIII-2. This difference may result from more frequent episodes of hearth maintenance during the level VIII-2 occupation. Endscrapers are concentrated in dumps in both levels; cores and burins both occur most often in functionally specific (southern) deposits. The two archaeological levels in Couche VIII have already been shown to be remarkably similar in overall assemblage composition, attributes of artifact form, technology, and use of the site locale. That the organization of space and the probable range of activities performed is also similar is not too surprising.

Differences between the occupations in aspects of spatial organization seem to relate primarily to different "intensities" of space use. More materials were

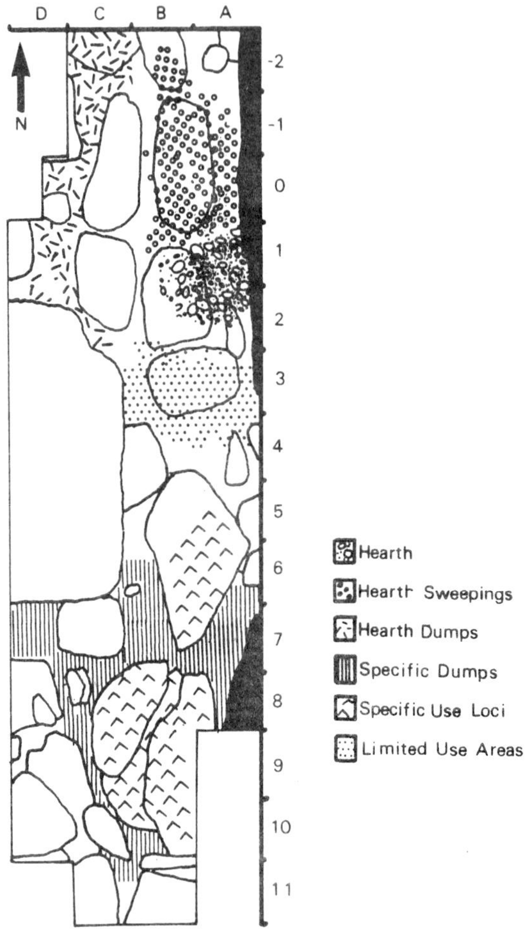

Figure 113. Hypothetical Reconstruction of Level VIII-2 Organization Based on Complete Spatial Analysis.

TABLE 51.

Percentages of artifacts by class contained in each inferred deposit type for Le Flageolet I level VIII-2.

ARTIFACT CLASS	DEPOSIT TYPE HEARTH	SWEEPINGS	DUMPS	SPECIFIC	SPARSE
Cores	12.5	12.5	29.2	37.5	8.3
Cobbles	28.6	51.2	6.6	13.6	0.0
Truncations	27.3	22.7	22.7	22.7	4.5
Burins	17.9	20.5	17.5	33.3	10.3
Endscrapers	20.6	20.6	38.2	20.6	0.0
Aurignacian	10.5	26.3	31.6	36.6	0.0
Retouched	0.0	65.2	17.4	17.4	0.0
Utilized	0.0	47.9	31.0	18.3	2.8
Notched	0.0	30.8	30.8	38.5	0.0
Debitage	14.0	29.1	20.4	36.5	0.0

abandoned, more dumps created, and more hearth maintenance deposits generated during the level VIII-2 occupation. Because only a single focal hearth occurs in each level, the variation cannot be explained by inferring the presence of two groups in level VIII-2 and only one in level VIII-1. Given a generally similar context of activity performance and site spatial organization, differences in intensity between the levels must be attributed to duration of occupation. Hearth rejuvenation and possible use of a secondary fire feature in level VIII-2 support the inference that it was a longer occupation than level VIII-1. This is not to imply that either occupation was very long. The surfaces hold little material compared to many Paleolithic sites (e.g., Pincevent), and few tools were abandoned during either occupation.

In sum, a redundant use of Le Flageolet is indicated by spatial analysis of two Aurignacian occupation levels from Couche VIII. While the exact nature of these occupations (i.e., what the people did there) is not inferred from these analyses, it does not seem likely that the shelter served as a semi-permanent habitation locus. Rather, the site was probably used by a single, small group for short periods of time in order to perform a particular set of activities using the flat blocks in the shelter's southern sector. More mundane group maintenance tasks were carried out in the northern sector around a single hearth. The redundant nature of occupation suggests that the activities performed at Le Flageolet were a regular part of a wider-scale regional Aurignacian subsistence system. However, the site and its assemblages cannot be viewed as representative of that wider system.

CHAPTER 9

SUMMARY AND DISCUSSION

INTRODUCTION

In concluding this investigation of spatial processes in Upper Paleolithic sites, two areas of interest will be discussed. First, the findings for each individual distribution will be summarized and compared. The purpose of this comparison is to determine if variability within the Aurignacian from Le Flageolet I can contribute to a broader understanding of (i) site use over time and (ii) each occupation's place in a regional system of economic behaviour. Pincevent represents a case study out of the temporal and regional context of Le Flageolet and will provide a point of contrast for the occupations of central concern.

The second issue to be addressed in these concluding remarks is more general. The efficacy of an heuristic approach to spatial analysis generally, and of the *k*-means technique specifically, will be assessed. Weaknesses in the preceeding analyses and how these can be addressed by future research will be outlined. Of major importance will be an assessment of basic assumptions concerning the nature of archaeological distributions, their meaning, and how to approach their analysis. The implications of the preceeding studies are important for the study of hunter-gatherer sites in general and for Paleolithic systematics in particular. These implications will also be discussed in this final chapter.

RESEARCH FINDINGS AND IMPLICATIONS FOR UNDERSTANDING PALEOLITHIC SITE USE

The preceeding analyses center on the recognition of spatial patterns in artifact distributions at several scales and the identification of depositional processes reflected in patterns of artifact class associations over space. The first case study involves the Magdalenian site at Pincevent excavated and analyzed by A. Leroi-Gourhan and M. Brézillon. Pincevent serves both as a test of the method to be applied to other surfaces and as a more specific study of spatial organization and site use. Patterning is defined at several levels of resolution. The organization detected both replicates and extends the informal interpretations of the original analysts.

Three large areas of material accumulation are identified at Pincevent based on k-means cluster analysis of ten artifact class distributions. These areas correspond almost exactly to the "habitation units" inferred by the excavators. Content analysis of these units shows that they are structurally similar (as is expected if three similar social units camped at the location and performed similar group maintenance activities). Spatial analyses also define smaller scale patterns in the Pincevent distributions. Some of the cluster zones defined here were recognized by Leroi-Gourhan and Brézillon, but some were not. Within each of the three large spatial units, refuse accumulations and smaller peripheral accumulations are present. Content analysis of these units demonstrated their homogeneous nature in all cases where stochastic effects can be ruled out. Based on these results, Pincevent is interpreted as a site where organized refuse disposal was an important depositional mode. This process caused materials generated in specific tasks to be deposited away from the locus of activity performance and associated with functionally unrelated materials. The explanation of this depositional process may be that contextual constraints on movement of refuse materials did not exist and/or that the presence of several social units constrained the mode of refuse disposal.

The Aurignacian occupations from Le Flageolet I are structurally quite different from Pincevent but are similar between themselves. The older level VIII-2 shows a two-sector division of space as. In the northern sector, a complex of deposits are associated with a focal hearth feature. Northern sector materials seem to relate to this hearth; some accumulations may reflect artifacts deposited by cleaning out the feature, while others represent refuse generated at the hearth and disposed of elsewhere. Some of the activities performed in the hearth context can be identified. It seems that lithic reduction occurred there since the densest concentrations of dechet and debitage are in hearth maintenance deposits. Expedient tool forms are concentrated around the focal feature, and endscrapers are also frequent in hearth area dumps. The southern sector of level VIII-2 is different than the north. The configuration of accumulations is not linear. A single rich concentration of materials occurs at the northern end of a large, flat rock, and two sparse deposits are present to the south. The assemblage profiles of all three southern accumulations are similar and quite distinct from northern concentrations. Burins and cores are the dominant elements of level VIII-2 southern accumulations. Utilized pieces are also present in the southern sector.

The uppermost Aurignacian surface at Le Flageolet I, level VIII-1, is very similar in organization to the level VIII-2 deposits. The basic dichotomy between northern and southern sectors holds again. A hearth is present in the level VIII-1 northern sector in the same place as in level VIII-2. However, the level VIII-1 focal hearth is small and diffuse and does not have clearly delimited maintenance deposits around it. A series of spatially discrete accumulations are associated with the northern feature. The assemblage profiles for these are very similar to the northern sector in the older Aurignacian level. Expedient tools are most frequent, and endscrapers are well-represented in nearby dumps. In the southern sector of level VIII-1, a linear pattern of accumulations follows the contours of the large southern block, and these are very similar in content. Again, southern sector zones have different assemblage profiles than northern accumulations. Burins and cores dominate southern sector assemblages.

Spatial organization in both Aurignacian levels from Le Flageolet I have been interpreted in the same way. Northern sectors of the occupations are inferred core areas where group maintenance activities occurred around a focal hearth feature. Activities in the core included lithic reduction and tasks involving the use of expedient tools and endscrapers. Other tools were also deposited into the core area but do not dominate the assemblages there. The presence of single hearth features in levels VIII-1 and VIII-2 indicate occupation by single groups. Neither the nature of those groups (families? task groups?) nor the number of people present is inferred. The richness of level VIII-2 dump deposits and the evidence for several episodes of hearth rejuvenation suggest that the duration of occupation was greater than for level VIII-1. Neither of the Aurignacian occupations was long, however.

The southern sector of both surfaces is interpreted as an "extensive use space" (Binford 1983) where a limited set of tasks was performed. These tasks included lithic reduction in both occupations. Burins were used during the level VIII-1 occupation, but truncations dominate in level VIII-2. The "activity areas" in the southern sector were probably located on the flat rock surface and may have involved the use of a small fireplace on that platform. Refuse from southern activities was disposed of into cracks around the block. In level VIII-1, linear patterns of refuse piles within the cracks suggest that activity performance involved multiple events, each of which generated small amounts of material. A single, dense deposit in level VIII-2 may represent disposal of refuse from many events into a single location. Thus, while disposal modes in the southern sector may have varied slightly between occupations, the use of the space did not.

Overall similarity in spatial organization during two Aurignacian occupations at Le Flageolet suggests that the site was used in the same way within a regional settlement system. Binford (1978b, 1983) has called such repeated use of a place "pattern redundancy" and relates it to logistically organized economic systems. In such systems, behaviours that on a regional scale provide a hunting group's subsistence are divided into behavioural elements that are "mapped" onto the environment. Each element relates a given place to important resources available near and at the location. Exploitation of those localized resources is planned so that the group carrying out procurement, the equipment taken to perform the necessary tasks, and the choice of places to "occupy" are suited to the resource extraction task at hand. When all elements are taken together, they compose an economic system reflected by a patterned yet variable archaeological record.

During the Aurignacian, Le Flageolet seems to have been used as such an element. The occupying groups were always small. The activities to be performed may have constrained the choice of place since they seem to have required the flat block surfaces present in Le Flageolet. While at the site, locally available foodstuffs were exploited, but it is not clear that this was the primary reason for occupation. The Aurignacian peoples arrived with tools in hand and left with tools they had made while at the site. In short, the rockshelter at Le Flageolet seems to have been used sporadically over a long period of time by small, mobile groups performing specific tasks. This implies that the assemblages from Le Flageolet are not representative of the full range of behaviours involved in Aurignacian subsistence and that the nature of occupation may not inform on overall characteristics of Aurignacian social relations.

The implications of logistical organization for traditional Paleolithic systematics are important. The stone tool assemblages left in a functionally specific location surely do not reflect, in the relative frequencies of individual artifact types, the "culture" of the people occupying the site. Without identifying all of the "places" composing the archaeological record of a single social/subsistence system, "cultural" or even economic relations among sites and regions cannot be examined. We simply do not know if we are examining parts of the same or a different system.

For traditional systemtics, serious questions are raised concerning the meaning of a given assemblage in regional chronologies. The classic approach to Paleolithic assemblage analysis views each site as a static and individual entity. Variation among these entities is considered "real" and meaningful in terms of the people occupying the sites. However, the archaeological record from an individual site may at best represent only one material "pose" assumed by a mobile group of people. Thus, the static view is limited and inappropriate if the goal of analysis is an understanding of past human behaviour. Furthermore, the behavioural meaning of traditional chronological facies and traditions is unknown. We must seek to understand this variability before inferring Paleolithic cultural relations.

EXPLAINING DIFFERENCES IN FORMATION PROCESS BETWEEN PINCEVENT AND LE FLAGEOLET

Although the Aurignacian occupations from Le Flageolet show remarkable similarity in nearly all aspects examined, there is an evident difference between the inferred site formation processes at Pincevent and Le Flageolet. At Pincevent, homogeneity in zone content does not decrease with spatial refinement. Accumulations become more heterogeneous as finer scale spatial divisions are made at Le Flageolet. Figure 114 overlays the several homogeneity graphs presented separately in each case study. The contrast between the three Le Flageolet occupations and Pincevent 36 is clear. As has been discussed, the Pincevent plot conforms to that predicted for a site where systematic refuse disposal changes the location of material discard from the location of generation. The plots for Le Flageolet are those anticipated when refuse from specific tasks is deposited in discrete spatial accumulations, i.e., is not mixed with debris from unrelated activities. Here, the locations of generation and discard are the same. Why are the sites distinct in disposal modes? In truth, the explanation of these differences is complex and cannot firmly be resolved at this stage of research.

Some ideas can be offered. Differences may be due to the occupied places. Pincevent is an "unbounded plain" in that movement across the surface was not hampered by topography. The collection and disposal of refuse may be a normal facet of any occupation, but it may have been constrained in performance by rocks at Le Flageolet. Thus, more discrete activity disposal may have been caused by the nature of the Le Flageolet space.

A second explanation for the differences may be the number of groups occupying the sites in question. Only a single group is inferred for each Le Flageolet occupation. Three "habitation units" are inferred for Pincevent, although only two of these were evidently contemporary (Leroi-Gourhan and Brézillon 1972). The several groups at Pincevent may have spaced themselves and disposed of refuse in ways a single group would not precisely because there were several groups present. At Le Flageolet we may see the "usual" record of a single social unit unconstrained by inter-group contact.

A final explanation is temporal. A great deal of time separates the Aurignacian from the Magdalenian. It has been widely argued that complexity of Magdalenian assemblages reflects the evolution of certain social

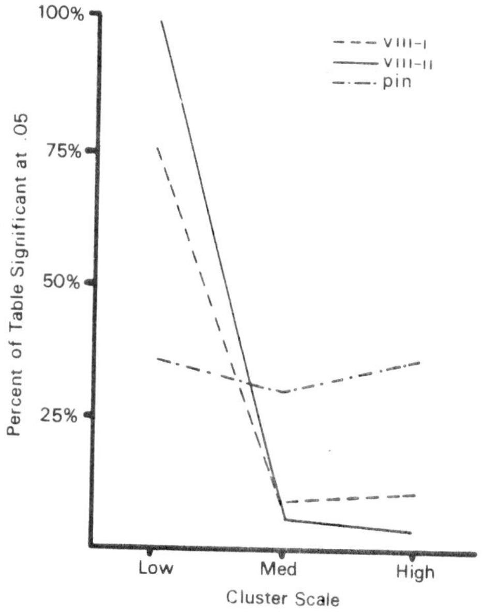

Figure 114. Homogeneity Plots for Analysed Levels Overlaid on a Single Graph.

behaviours in response to increasing population and inter-group dynamics at the end of the Pleistocene (e.g., Bahn 1977; Conkey 1978, 1980, 1983). This explanation has been invoked to account for the "explosion" of paleolithic art and increased number of sites attributed to the Magdalenian.

Magdalenian sites in the Perigord are the most numerous, and (with a few exceptions) their rockshelters are the largest (White 1980). Open air Magdalenian sites are also present in the region, and these tend to comprise several habitations indicating the presence of several groups (Gaussen 1980). In a context of increasing social interaction, refuse management may have developed in response to multi-group settlements and a correlated need to maintain socially defined spaces within an occupation. An "explosion of garbage dumps" coinciding with other apparant social changes is an intriguing possibility. However, much more research needs to be done before such an explanation could be supported.

The present analyses deal with occupations in only two places at the extremes of Upper Paleolithic times. The contexts of the two sites are different. While the consistent use of Le Flageolet over time is demonstrated, evolutionary implications of differences between Pincevent and Le Flageolet must await a more complete sequence of similarly analyzed levels. At present, a simple contextual (i.e., topographic) explanation seems most reasonable.

ASSESSMENT OF THE METHODS AND THEIR IMPLICATIONS

The *k*-means procedure has proved an effective method for analyzing archaeological spatial distributions. Several characteristics of the technique are of greatest importance. First, the program's ability to detect patterns at many scales and to identify appropriate pattern solutions is unique among available methods. In all studies presented here, large scale patterning, obvious to the naked eye, is present in the distributions. Finer scale patterns are also detected that are not apparant using informal techniques. For example, the excavators of Pincevent noted and interpreted the three large scale accumulations but were less precise in distinguishing small deposits. The ability of the human eye to see patterns *within* other patterns is variable. Other formal methods applied to archaeological distributions have ignored the possibility of this kind of multiple pattern recognition (cf. Simek and Larick 1983).

Many of the formal methods previously used in archaeology lack the means to compare patterns among maps. I have argued that this is due, in part, to the reductive nature of these techniques. Often, the spatial relevance of any detected patterning is lost as distributions are reduced to single descriptive statistics. Characterizations such as "random" or "clustered" cannot be translated back to the site map for examination and comparison. Even when artifact classes are analyzed for association, there is no way to tell where on a surface such association occurs. Methods performing this kind of data reduction are not very useful for addressing archaeological problems. The *k*-means technique, by providing locational information and data on cluster size, allows optimal solutions for a variety of class distributions to be compared. The retention of locational information also allows the integration of patterns in artifact distributions with contextual data such as the locations of hearths and local topography.

While the utility of *k*-means cluster analysis as a pattern recognition technique is established, other methods used in the preceeding analyses are less satisfying. Although the heuristic definition of cluster zones and the comparison of their contents is performed with successful results, more rigorous and formalized techniques for carrying out these tasks are desirable. In particular, a more direct measure of homogeneity is needed, perhaps one examining the relation between sample size and diversity of accumulation contents. Two possible approaches to this problem have been proposed recently (Boone 1982; Kintigh 1982, 1984). Both techniques employ modelled expectations for diversity in assemblage content under random sampling processes. These methods suggest directions for future development of ways to assess homogeneity, and these directions are currently being pursued (e.g., Larick 1983; Simek 1984; Simek, *et al*. 1984).

Less informal methods for defining zones are also a goal of continuing research. One approach to this problem was outlined in Chapter 3 and uses the centroid and RMS statistics produced by *k*-means to assess cluster coincidence in space. As was discussed earlier, the nature of RMS definition must be understood better than it is now for such an approach to be fruitful. Perhaps by incorporating the density of materials composing a cluster, i.e., the point frequency, RMS radii can be weighted to express the actual dispersal of items associated in a cluster. Quantitative assessments of RMS overlap might then be more relevant to the problem. This is a technical issue that will also be pursued in future research.

In all, the techniques employed here for the analysis and comparison of archaeological distributions have succeeded in providing insights into possible site formation processes at Pincevent and Le Flageolet I. Spatial accumulations have been defined and related to multiple site formation processes. These were compared within a single distribution and between distributions.

In short, the techniques have accomplished the goals outlined in Chapter 3 and have allowed different kinds and sizes of patterns to be defined and compared.

GENERAL IMPLICATIONS AND CONTRIBUTIONS OF THE ANALYSES

Results of the foregoing studies have general implications for examining spatial distributions from hunter-gatherer archaeological sites. Perhaps the most important implication concerns traditional assumptions about the kinds of information such distributions hold. Archaeologists have usually assumed that spatial patterns in artifact distributions reflect the locations of activities directly (or can be coaxed into reflecting activity areas if the proper "transforms" are posited; e.g., Schiffer 1976). The studies presented here show that spatial patterns cannot presumptively be related to specific behaviours but must be viewed in terms of depositional structure. After all, archaeological materials arrive in the record by deposition. To explain them by reference to behaviours that cannot be empirically verified instead of as varieties of deposit than can be observed, is to ignore Binford's reminder that the archaeological record is a static and contemporary fact (e.g., 1983). The empirical record is a complicated entity that must be understood first in its own terms. Then, perhaps, the task of relating the static facts to past dynamic generative processes can be addressed.

The research presented here has attempted to account for that complexity. The goal of descriptive analysis was to identify varieties of accumulations over space. These are viewed as having many potential causes, and models for explaining observed variability were proposed that incorporate multiple formation processes. That variation exists in the kind of deposits over a single archaeological surface and that such variation can be defined analytically, is one of the principal contributions made in this work. Clearly, we can no longer approach the archaeological record assuming that all spatial distributions will reveal activity areas.

Another contribution of the analyses presented here concerns the utility of reconstructions based on spatial analysis. When the goal of study is shifted from identifying past activities to describing the present record, site comparisons and classifications become easier and more interesting in many respects. The inferences drawn here for the surfaces under study are explicitly organizational rather than reconstructive. What the Aurignacian inhabitants of Le Flageolet were doing is not inferred. That their use of the location was redundant and locally intensive over a long period of time is important information when considering land use and regional settlement systems. The Le Flageolet analyses suggest that Aurignacian land use was mobile and logistically organized and that traditional interpretations for similarities and differences among sites may be inaccurate. To identify all the different kinds of sites (in an organizational sense) that compose Aurignacian settlement, we do not need to know what their specific activities were.

The traditional goals of spatial analysis in archaeology, i.e., the definition of site types for integration into regional land use models, need not require reconstruction at all. In fact, evolutionary explanations of changes in settlement patterns may be made more difficult by reconstruction that involves "tool kits" and "activity areas". These notions emphasize static formulations rather than the dynamic nature of human responses to selective pressures. Individual sites contain the data that must be analyzed to address the problem of adaptation, but selection and change cannot be viewed in a single occupation. Reconstructing what went on during a single occupation at a single location may expend resources to obtain uneeded information. In order to understand human adaptation to a given environment, and changes in adaptation over time, we need to know how many different kinds of places composed an ongoing adaptive system, how each place fits into the system of all places, and how that system related to stability and change in the environment. We may not need to know how a particular item was used since specific uses are almost random events, influenced by a multitude of factors, that cannot ultimately be verified in systematic tests.

CONCLUDING REMARKS

In closing, some directions for further research suggested by these studies can be proposed. In addition to the technical work remaining to be done, identifying causes of variation between Pincevent and Le Flageolet poses an interesting problem. Differences between the distributions from the two sites might be due to topographic factors, social relations, functional variability, and/or evolutionary change. To reduce this somewhat bewildering array of potential explanations, research must concentrate on (i) refining the variables analyzed so that measurement can be related directly to causal process (i.e., "Middle Range Theory") and (ii) including a more complete series of cases in the equation that provides control over time, context, site size, site use, etc.

As a first step, the later archaeological occupations at Le Flageolet I (the Upper Perigordian levels) will be studied. These levels will provide information from time periods between the Aurignacian and Magdalenian and allow the "evolutionary" explanation to be assessed. They will also provide information concerning the effects of topography on spatial organization within Le Flageolet, since the upper layers are not influenced by large blocks as are the Aurignacian levels.

Analysis of the level IX and XI distributions after complete excavation will allow the extent of occupation redundancy to be assessed and will allow the interpretive framework used here to be tested. Such studies will provide a direct test of the assumed structural relations between artifacts and contextual elements. Spatial analysis of bone data from all levels, once identification of species and body part is complete, will also provide empirical tests for some of the inferences made here. These data were collected with point provenience, and they can be added to all stages of spatial and content analysis.

These are a few of the avenues for further study suggested by the results at hand. They will be pursued. It is hoped, moreover, that this work has pointed to some more general dimensions of potential variation in hunter-gatherer sites that must be taken into account in any spatial analysis. We cannot continue to choose our methods based on false assumptions concerning what the record is and holds.

Finally, it is hoped that the results achieved here concerning specific Upper Paleolithic sites will be of use to scholars interested in the archaeology of that time period. It has been demonstrated that all of the analyzed levels represent small scale occupations within larger settlement systems. Moreover, they are different from one location to another. Both interpretations imply that other loci must be sought if large scale "base camps" are to be found. It may be that sites such as Pincevent and Le Flageolet are typical, and that the kinds of permanent, large scale settlements assumed to characterize Paleolithic systems do not exist. These results also imply that site-specific "cultural" identifications are meaningless for wide ranging, logistically organized economic systems.

If these limited conclusions provide even slight encouragement for others to pursue (and modify or falsify) the concerns at the hearth of this study, then it will have been successful.

REFERENCES CITED

Ammerman, A. and M. Feldman

1974 On the making of an assemblage of stone tools. _American Antiquity_ 39(4):610-616.

Bahn, P.

1977 Seasonal migration in southwest France during the Late Glacial period. _Journal of Archaeological Science_ 4:245-57.

Binford, L.

1973 Interassemblage variability--the Mousterian and the "functional" argument. In _The Explanation of Prehistoric Change_, edited by C. Renfrew. London: Duckworth. Pp. 227-54.

1976 Forty-seven trips: a case study in the character of some formation processes. In Contributions to Anthropology, the Interior Peoples of Northern Alaska, edited by E. Hall. _National Museum of Man, Mercury Series No. 49_. Ottawa: National Museum of Canada. Pp. 299-351.

1977 General Introduction. In _For Theory Building in Archaeology_, edited by L. Binford. New York: Academic Press. Pp. 1-10.

1978a Dimensional analysis of behaviour and site structure: learning from an Eskimo hunting stand. _American Antiquity_ 43:330-61.

1978b _Nunamiut Ethnoarchaeology_. New York: Academic Press.

1979a Comments on confusion. _American Antiquity_ 44(4):591-94.

1979b Organization and formation processes: looking at curated technologies. _Journal of Anthropological Research_ 35:255-73.

1981a Behavioural archaeology and the "Pompeii" premise. _Journal of Anthropological Research_ 37:195-208.

1981b _Bones: Ancient Men and Modern Myths_. New York: Academic Press.

1983 In Pursuit of the Past. London: Thames and Hudson.

Binford, L. and S. Binford

1966 A preliminary analysis of functional variability in the Mousterian of Levallois facies. American Anthropologist 68(2):238-95.

Blalock, H.

1972 Social Statistics. New York: McGraw-Hill.

Boone, J.

1982 Defining and measuring refuse catchment. Paper presented at the 47th Annual Meeting of the Society for American Archaeology, Minneapolis.

Bordes, F.

1978 Le Protomagdalénien de Laugerie-Haute Est (fouilles Bordes). Bulletin de la Société Préhistorique Française 72:139-144.

Bordes, F. and J. Labrot

1967 La stratigraphie du gisement du Roc de Combe (Lot) et ses implications. Bulletin de la Société Préhistorique Française 64:29-34.

Clark, G.

1979 Spatial association at Liencres, an Early Holocene open site on the Santander Coast, North-central Spain. In Computer Graphics in Archaeology: Statistical Cartographic Applications to Spatial Analysis in Archaeological Contexts, edited by S. Upham. Arizona State University Anthropological Research Papers No. 15. Tempe: Arizona State University Press.

Clarke, D.

1968 Analytic Archaeology. London: Methuen.

1973 Archaeology: the loss of innocence. Antiquity 47:6-18.

1977 Spatial information in archaeology. In Spatial Archaeology, edited by D. Clarke. New York: Academic Press. Pp. 1-32.

Collingwood, R.

1946 *The Idea of History*. Oxford: Oxford University Press.

Conkey, M.

1978 Style and information in cultural evolution: toward a predictive model for the paleolithic. In *Social Archaeology: Beyond Subsistence and Dating*, edited by C. Redman, M. Berman, E. Curtin, W. Langhorne, N. Versaggi, and J. Wanser. New York: Academic Press. Pp. 61-85.

1980 Context, structure, and efficacy in paleolithic art and design. In *Symbol as Sense*, edited by M. Foster and S. Brandes. New York: Academic Press. Pp. 225-48.

1983 On the origins of paleolithic art: a review and some critical thoughts. In The Mousterian Legacy: Human Biocultural Change in the Upper Pleistocene, edited by E. Trinkhaus. *BAR International Series 164*. Oxford: BAR. Pp. 201-27.

Cowgill, G.

1968 Archaeological applications of factor, cluster, and proximity analysis. *American Antiquity* 33(3):367-75.

Dacey, M.

1973 Statistical tests of spatial association in the locations of tool types. *American Antiquity* 38:320-28.

Delpech, F.

1970 L'abri magdalénien du Flageolet II (Bézenac, Dordogne)-- Paléontologie. *Bulletin de la Société Préhistorique Française* 67:494-99.

1975 *Les Faunes du Paléolithique Supérieur dans le Sud-Ouest de la France*. Thèse de Doctorat d'État ès Sciences Naturelles. Université de Bordeaux I. No. 479.

Delpech, F. and J-Ph. Rigaud

1974 Étude de la fragmentation et de la répartition des restes osseux dans un niveau d'habitat paléolithique. In *Premier Colloque International sur l'Industrie de l'Os dans la Préhistoire*, edited by H. Camps-Fabrer. Aix-en-Provence: Université de Provence. Pp. 47-55.

Delporte, H.

1968 L'Abri Facteur à Tursac (Dordogne): I etude générale. _Gallia Préhistoire_ 11:1-112.

Delporte, H. and G. Mazière

1977 L'Aurignacien de la Ferrasie: observations préliminaires de la suite des fouilles récentes. _Bulletin de la Société Préhistorique Française_ 74:343-61.

Diot, M-F.

1981 Report to Jean-Philippe Rigaud on pollen from Le Flageolet I. Manuscript on file at the Direction des Antiquites Prehistoriques d'Aquitaine, Bordeaux.

Doran, J. and F. Hodson

1972 _Mathematics and Computers in Archaeology_. Cambridge: Harvard University Press.

1975 _Mathematics and Computers in Archaeology_, 2nd Edition. Cambridge: Harvard University Press.

Fenelon, P.

1951 _Le Périgord: Étude Morphologique_. Paris: Lahure.

Freeman, L.

1978 The analysis of some occupation floor distributions from earlier and middle Pleistocene sites in Spain. In _Views of the Past_, edited by L. Freeman. The Hague: Mouton. Pp. 57-116.

Gaussen, J.

1980 Le Paléolithique Supérieur de Plein Air en Périgord: Secteur Mussidan--Saint Astier--Moyenne Vallée de l'Isle. _16eme Supplément a Gallia Préhistoire_. Paris: C.N.R.S.

Gould, R.

1977 _Living Archaeology_. New York: Cambridge University Press.

Graham, I.

1980 Spectral analysis and distance methods in the study of archaeological distributions. Journal of Archaeological Science 7:105-29.

Hietala, H. and D. Stevens

1977 Spatial analysis: multiple procedures in pattern recognition studies. American Antiquity 42:539-59.

Hodder, I. and E. Okell

1978 An index for assessing the association between distributions of points in archaeology. In Simulation Studies in Archaeology, edited by I. Hodder. New York: Cambridge University Press. Pp. 97-108.

Hodder, I. and C. Orton

1976 Spatial Analysis in Archaeology. New York: Cambridge University Press.

Hodson, F.

1970 Cluster analysis in archaeology: some new developments and applications. World Archaeology 1(3):299-320.

1971 Numerical typology and prehistoric archaeology. In Mathematics in the Archaeological and Historical Sciences, edited by F. Hodson, D. Kendell, and P. Tautu. Edinburgh: Edinburgh University Press. Pp. 30-45.

Johnson, A. and A. Johnson

1975 K-means and temporal variability in Kansas City Hopewell ceramics. American Antiquity 40:283-95.

Johnson, I.

1976 Contribution Méthodologique à l'Étude de la Répartition des Vestiges dans des Niveaux Archaeologiques. Diplome des Études Supérieur, Université de Bordeaux I.

Keeley, L.

1980 Experimental Determination of Stone Tool Uses: A Microwear Analysis. Chicago: University of Chicago Press.

1981 Hafting and retooling: effects on the archaeological record. American Antiquity 47:798-809.

Kempthorne, O.

1969 An Introduction to Genetic Statistics. Ames: Iowa State University Press.

Kershaw, K.

1964 Quantitative and Dynamic Ecology. London: Edward Arnold.

Kintigh, K.

1982 A painless measure of archaeological diversity. Paper presented at the 47th Annual Meeting of the Society for American Archaeology, Minneapolis.

1984 Measuring archaeological diversity by comparison with simulated assemblages. American Antiquity 47(1):31-63.

Kintigh, K. and A. Ammerman

1982 Heuristic approaches to spatial analysis in archaeology. American Antiquity 47(1):31-63.

Larick, R.

1983 The Circulation of Solutrean Foliate Point Cherts: Residential Mobility in the Perigord. Unpublished Ph.D. Dissertation, State University of New York at Binghamton.

Laville, H.

1970 L'abri magdalénien du Flageolet II (Bézenac, Dordogne)-- étude géologique. Bulletin de la Société Préhistorique Française 67:475-488.

1975 Climatologie et Chronologie du Paléolithique en Périgord: Étude Sédimentologique de Depots en Grottes et sous Abris. Aix-en-Provence: Université de Provence.

Laville, H. and J-Ph. Rigaud

1974 The Perigordian V industries in Perigord: typological variations, stratigraphy and relative chronology. *World Archaeology* 4:330-38.

Laville, H., J-Ph. Rigaud and J. Sackett

1980 *Rockshelters of the Perigord: Geological Stratigraphy and Archaeological Succession*. New York: Academic Press.

Leroi-Gourhan, A. and M. Brézillon

1966 L'habitation magdalénienne no. 1 de Pincevent près de Montereau (Seine-et-Marne). *Gallia Préhistoire* 9(2):263-385.

1972 Fouilles de Pincevent: Essai d'Analyse Ethnographique d'un Habitat Magdalénien. *7eme Supplément a Gallia Préhistoire*. Paris: C.N.R.S.

Movius, H. and A. Brooks

1971 The analysis of certain major classes of Upper Paleolithic tools: Aurignacian endscrapers. *Proceedings of the Prehistoric Society* 37(2):253-73.

Movius, H., N. David, H. Bricker, and B. Clay

1968 *The Analysis of Certain Major Classes of Upper Paleolithic Tools*. Cambridge: Peabody Museum Press.

Odell, G. and F. Odell-Vereecken

1980 Verifying the reliability of lithic use-wear assessments by "blind tests": the low power approach. *Journal of Field Archaeology* 7:87-120.

Pielou, E.

1977 *Mathematical Ecology*. New York: John Wiley and Sons.

Pinder, D., I. Shimada, and D. Gregory

1979 The nearest neighbor statistic: archaeological applications and new developments. *American Antiquity* 44(3):430-445.

Rigaud, J-Ph.

1969 Note préliminaire de la stratigraphie du gisement du "Flageolet I" (commune de Bézenac, Dordogne). _Bulletin de la Société Préhistorique Française_ 3:73-5.

1970 Étude préliminaire des industries magdaléniennes de l'abri du Flageolet II (commune de Bézenac, Dordogne). _Bulletin de la Société Préhistorique Française_ 67(2):456-74.

1976a Les civilizations du paléolithique supérieur en Périgord. In _La Préhistoire Française_, edited by H. de Lumley. Paris: C.N.R.S.

1976b Les gisements de Flageolet, commune de Bézenac. In _Livret Guide A4: Sud-ouest (Aquitaine et Charante)_, edited by B. Vandermeersh and J-Ph. Rigaud. Nice: UISPP.

1978a The significance of variability among lithic artifacts: a specific case from southwestern France. _Journal of Anthropological Research_ 34(3):299-310.

1980 Données nouvelles sur l'Aurignacien et le Périgordien en Périgord. In _Colloque International sur l'Aurignacien et le Gravettien (Périgordien) dans leur Cadre Écologique_. Cracow and Nitra: UISPP.

1982 _Le Paléolithique en Périgord: Les Données du Sud-Ouest Sarladais et leur Implications_. Thèse de Doctorat d'État ès Sciences Naturelles. Université de Bordeaux I. No. 737.

Robbins, G., N. Seeley, M. Symons, and D. McNeil

1978 Identification of ancient heat treatment in flint artifacts by ESR spectroscopy. _Nature_ 276:703-4.

Rummell, R.

1970 _Applied Factor Analysis_. Evanston: Northwestern University Press.

Sackett, J.

1966 Quantitative analysis of Upper Paleolithic stone tools. _American Anthropologist_ 68(2):356-94.

Scargill, I.

1974 The Dordogne Region of France. North Pomfret, Vt.: David and Charles.

Schiffer, M.

1976 Behavioral Archaeology. New York: Academic Press.

Simek, J.

1981 The structure of space in a rockshelter: Le Flageolet I level VIII-superieur. Paper presented at the 46th Annual Meeting of the Society for American Archaeology, San Diego.

1984 Structure and diversity in intrasite spatial analysis. Paper presented at the 49th Annual meeting of the Society for American Archaeology, Portland, Oregon.

Simek, J. and R. Larick

1983 The recognition of multiple spatial patterns: a case study from the French Upper Paleolithic. Journal of Archaeological Science 10:165-80.

Simek, J. and P. Leslie

1983 Partitioning chi-square for the analysis of frequency table data: an archaeological application. Journal of Archaeological Science 10:79-85.

Siegal, F.

1956 Nonparametric Statistics. New York: McGraw-Hill.

Skoog, R.

1968 Ecology of the Caribou in Alaska. Unpublished Ph.D. Dissertation, University of California.

Sneath, P. and R. Sokol

1976 Numerical Taxonomy. San Francisco: W. H. Freeman.

Sonneville-Bordes, D. de and J. Perrot

1954-56 Lexique typologique du paléolithique supérieur. Bulletin de la Société Préhistorique Française (51):327-35; (52):76-9; (53):408-12, 547-59.

Speth, J. and G. Johnson

1976 Problems in the use of correlation for the investigation of tool kits and activity areas. In Cultural Change and Continuity, edited by C. Cleland. New York: Academic Press. Pp. 35-57.

Texier, J-P.

1979 Recherches sur les Formations Superficielles du Bassin de l'Isle. Thèse de Doctorat d'État ès Sciences Naturelles. Université de Bordeaux I. No. 613.

Van Noten, F., D. Cahen, and L. Keeley

1979 Stone tools, toolkits, and human behaviour in prehistory. Current Anthropology 20:661-72.

Whallon, R.

1973 The spatial analysis of occupation floors I: application of dimensional analysis of variance. American Antiquity 38:320-8.

1974 The spatial analysis of occupation floors II: the application of nearest neighbor analysis. American Antiquity 39:16-34.

Wheat, J.

1973 The Olsen-Chubbuck Site: A Paleo-Indian Bison Kill. Society for American Archaeology Memoirs No. 29.

White, R.

1980 The Upper Paleolithic of the Perigord: A Topographic Approach to Subsistence and Settlement. Unpublished Ph.D. Dissertation, University of Toronto.

Wilmsen, E.

1970 Lithic Analysis and Cultural Inference: A Paleo-Indian Case. University of Arizona Anthropological Papers No 16. Tucson: University of Arizona Press.

1974 Lindenmeier: A Pleistocene Hunting Society. New York: Harper and Row.

1978 Lindenmeier. Washington, D.C.: Smithsonian Institution.

Wilson, J.

1975 The Glacial environment at the Abri Pataud. In Excavations at the Abri Pataud, Les Eyzies (Dordogne) Vol. 1, edited by H. Movius. Cambridge: Peabody Museum Press.

Yellen, J.

1977 Archaeological Approaches to the Present: Models for Reconstructing the Past. New York: Academic Press.